Geochemical Exploration and Modelling of Concealed Mineral Deposits

Ashoke K. Talapatra

Geochemical Exploration and Modelling of Concealed Mineral Deposits

 Springer

Ashoke K. Talapatra
Geological Survey of India
Kolkata, India

ISBN 978-3-030-48755-3   ISBN 978-3-030-48756-0   (eBook)
https://doi.org/10.1007/978-3-030-48756-0

Jointly published with Capital Publishing Company, New Delhi, India

The print edition is not for sale in SAARC countries. Customers from SAARC countries –
Afghanistan, Bangladesh, Bhutan, India, the Maldives, Nepal, Pakistan and Sri Lanka,
please order the book from:

Capital Publishing Company, 7/28, Mahaveer Street, Ansari Road, Daryaganj, New Delhi
110 002, India.

This Springer imprint is published by the registered company Springer Nature Switzerland AG
The registered company address is: Gewerbestrasse 11, 6330 Cham, Switzerland

Dedicated to
My beloved parents
Basanti and Aswini Kumar Talapatra
and
To my wife Sukla for her co-operation

Preface

Locating new economic mineral deposits is becoming increasingly difficult now in any country, since there is little chance for searching fresh *in situ* outcrops with surface signature of mineralisation that has not been geologically mapped or reported by national surveys or other agencies. This is because in a country like India the Geological Survey of India (GSI) which is one of the oldest organization of the world is actively engaged in geological work for about 170 years and the entire country has been mapped on 1 : 50,000 scale by now. Similarly most of the countries of the world have completed geological mapping of their respective countries by now. As such there is hardly any probability of reporting new terrestrial and offshore occurrences or deposits that have not been known till date by conventional methods of exploration. Surprisingly, our ancestors could locate in countries with ancient civilization like India and some other countries, number of iron, base-metal (Cu, Pb, Zn, etc.), gold, silver deposits in various parts, long long years ago, perhaps 2000 or more years earlier. In the context of India, certain parts of Rajasthan, Karnataka, Jharkhand, Orissa and some other states, very old mining activities could be located beyond 170 m depths without using any electricity or any mechanical device of modern age. The situation is somewhat different in the case of offshore mineral deposits which are generally present along the entire coastal belt up to the Exclusive Economic Zone (EEZ) of any country having large coastal areas, as a continuation of existing beach placers or outcrops of any mineral deposit like limestone etc. that continue from sea beach towards offshore. Ancient people used to ultilise beach placers and other onshore mineral deposits according to their requirements. In this respect, Marine Wing, Geological Survey of India (GSI) like many other countries has already scanned systematically almost entire EEZ of India by wide spaced traverses of different cruises (5 km or more) at the initial stage taking different types of samples with the help of its three research vessels (RV).

The only remaining prospects in the land area at present are those where deposits of any known metallogenic belt/province are concealed along the strike direction beneath the soil or other types of barren overburden/talus materials. Alternatively, there may be some virgin occurrences/deposits that are totally covered by such material in a favourable geological set-up. In this respect concept of crustal evolution and metallogeny is very important, especially after global recognition of the plate-tectonic hypothesis. Potential mineral belts in global perspective with special reference to Indian mineral belts in different geotectonic frame will be discussed here to suggest new mineral belts for geochemical exploration and modelling. Concealed terrestrial and offshore mineral deposits are discussed in this book with suggestions to find out such deposits in any part of the world. Such an approach will cater to the needs of explorers of mineral deposits and exploration geologists covering larger regions of the world, particularly the countries of South and South-East Asia. Moreover, students and research scholars of Economic Geology and Mineral Exploration are expected to be immensely benefited by this.

In this connection, I would like to stress upon the importance of Rare Earth Element (REE) and Rare Metal (RM) exploration compared with search of other common minerals in world perspective. In fact prior to year 2010, hardly any one even knew the most essential use of rare earths in day-to-day life. Subsequently, there has been a sudden explosion in demand of many essential items of use by mankind that require rare earths. Deposits of REE and RM are not so common within the different countries of the world, and these deposits are rarely found exposed along the surface. There are some possibilities of locating some concealed deposits which require specialized techniques of exploration that will be discussed in the book.

Coming to Indian perspective, beach sands along the east and west coasts of India were well known for their presence along the coastal areas since long. These hold almost 35% of the world's total beach sand mineral deposits, which are known to be significant source of rare earths of present day India. But the rare earths present within these heavy minerals of sea coast were not known to be so much useful in earlier days. Exploration of beach sands along the coastal areas of both east and west coasts of India has been discussed in details in the book along with specific techniques to be used. However, within the crustal areas of Indian subcontinent Precambrian pegmatitic rocks in association with carbonatite bodies, present along prominent lineaments, are expected to unearth significant deposits of REE in future. So, concealed deposits of REE and RM deposits in such terrains should be discovered applying new techniques of exploration.

In the case of offshore mineral deposits/occurrences (mostly of heavy mineral placers, phosphates, carbonates, sulphides, oxides, crude oil, gas hydrates etc.), systematic close spaced sampling along the offshore areas within the EEZ may locate some good deposits, provided the results of initial grab and other types of samples from the sea bed, wherever available, give some encouraging results. In fact a wide variety of mineral resources have been reported from the continental margin with ocean, some of which have been explored since ancient past. Besides, proving the presence of large amount of manganese nodules of various sizes in certain parts of deep ocean, another very important type of recent deposits are represented by hydrothermal mineralisation within the modern sea floor, along the spreading mid-oceanic ridges, as a consequence of plate movements. These are, however, beyond the scope of this work.

Under such conditions, be it for land deposit or offshore deposit, application of statistical methods with qualitative and, wherever possible, quantitative earth science data for evaluation of mineral resources has become common in recent times. In order to utilise the huge amount of data collected by the different government agencies and undertakings, computer-based mineral deposit modelling is the best suitable way out for predicting concealed mineralisation along the strike continuity of known deposits, or in the virgin areas, where geological set-up is somewhat similar, but without showing any surface indication of mineral occurrence. These types of concealed deposits will be illustrated in this book with some case histories. The essential pre-requisite for such an approach is the development of a full fledged database containing different types of earth science data, both alpha-numeric and graphic type. All the data pertaining to the area of study should be geo-referenced. That means each sample point from where different types of data will be generated should have accurate latitude (Y) , longitude (X) and difference of height from mean sea level (Z) values, wherever possible. These should be plotted with the help of a suitable GIS package so that they may be shown in a map. In the case of offshore data or data from drill cores/mine levels, 'Z' values with respect to mean sea level, should also be recorded.

It has been observed that the term "modelling" is being used loosely in geological literature and some times a mere geological account of an area is also considered as modelling (Sarkar and Rai, 2002). In fact, modelling is a representation of a body in three dimensions, which in the present context, is mineral deposit. Mineral deposit modelling, which is practised for more than two decades or so, for land-based deposits, may be of different types. Computerized approaches adopted for this will be discussed in this monogram. However, the main objective of modelling is to delineate the body geometry of the deposit in three dimensions so that the mineralised body or lode zones may be properly assessed and exploited by the scientific approaches of mining to the highest extent. In this respect, modelling may be used for offshore mineral deposits also, with suitable 3-D extensions, provided sufficient underwater data of the deposit are available. This will be illustrated in the text. In short, modelling of terrestrial and offshore mineral deposits replicates its body geometry along with its physical distribution of various parameters, in order to find out its 3-D configuration.

In order to assess the three dimensional configuration of the ore-body or lode zone for a land-based deposit, both surface and sub-surface data are required, and this will definitely help detailed exploration of the concealed deposit. Such mineral exploration should be initiated with special emphasis on exploration geochemistry, which includes both conventional and non-conventional techniques, which will be discussed in detail later. In this monogram, the author's endeavour has been to compile the different techniques of exploration geochemistry, giving more emphasis to concealed land-based mineral deposits of the different countries, that are covered by transported soil, sand, alluvium, talus material etc. During the last three decades, scientists of GSI could standardize some new techniques of exploration geochemistry, with indigenous equipments suitable in Indian condition. Various geophysical exploration techniques, employed in different parts of the earth, could locate a number of anomalous zones, some of which subsequently could prove the presence of concealed deposits. But mostly, these techniques are not able to ascertain the nature and composition of the causative bodies, or nature of the deposit, without the help of ground geological and geochemical surveys. Some of these geophysical anomaly zones, proved by subsequent drilling, appear to be due to the presence of graphitic rocks, brine water etc., instead of any economic ore deposit. In this respect, the geochemical surveys are advantageous, as these techniques can pinpoint the nature and type of element of interest concealed behind these anomalies, sometimes using the presence of its pathfinder elements in anomalous amounts. Standard analytical methods of geochemical samples have been outlined in short, so that geologists can suggest the appropriate method of analysis, to be adopted by the chemists, depending upon the type of sample and problem. This write-up is intended to serve as a guide to field geologists, geochemists, students, research scholars and scientists interested in earth science for exploration of mineral deposits in any part of the world and for evaluation of the resources.

In course of compiling this write-up, there may be some omissions. However, the author will feel amply rewarded, if this compilation assists the present and future geologists, students and research scholars of earth science, to plan their mineral exploration programme for exploiting concealed mineral deposits. Especially the hidden deposits, which are covered along the strike continuity of the known mineral belt or those which show similar geological set up, having identical structural history with some geostatistical characteristics of major and trace element distribution that are most important for detailed study. In case of offshore mineral deposits, 3-D modelling with vibrocore samples collected with the help of coastal vessels give sufficient indications of extension of placer deposits within the EEZ, which will be discussed in the book.

In fact, reserve of various important mineral commodities are depleting very fast. So, unless we can locate the presence of concealed deposits of the ore minerals like Cu, Pb, Zn, Ni, Ag, Au, U, Li, Ce etc., future generation will be facing lots of problems. However, application of Zipf's Law (1949), a mathematical relationship between size and rank of discrete phenomena, has been used for resource prediction of oil in Western Canada, uranium, gold, tin and lead-zinc deposits in Australia, and copper deposits in Zambia. In India, Paliwal et al. (1986) and Halder (2004) could establish the probable presence of concealed deposits of Pb, Zn etc. by studying twenty four known Precambrian sulfide deposits and according to them 75% of Pb-Zn metals are yet to be discovered. Application of this law is a very interesting observation, and may be applied for other commodities also. With the gradual depletion of various important mineral commodities, immediate attention of the earth scientists should be concentrated to try their best right now, to discover concealed deposits in their respective countries.

Before concluding the Preface of this book, let me again acknowledge the help and co-operation received by me for the book from all my well wishers, friends, colleagues and faculty members of Geology Dept., Presidency College, Kolkata and also my family members. I would also like to express my thanks to Prof. (Dr.) Pradip K. Sikdar, IISWBM, Kolkata and to Ms. Angana Chowdhury, Sr. Research

Scholar, Indian Institute of Technology, Bombay for helping me by sharing recent and relevant scientific publications on this subject for my use. Thanks are also due to my son Angshuman and my daughter Nilanjana who helped me in computer-based processing of the manuscript during their spare time.

Kolkata, India   Ashoke K. Talapatra
July 2019

References

Halder, S.K. (2004). Grade and tonnage relationships in sediment hosted lead zinc sulphide deposits of Rajasthan, India. *In:* Deb, M. and Goodfellow, W.D. (Eds.), Sediment hosted lead-zinc sulphide deposits: Attributes and models of some major deposits in India, Australia and Canada. Narosa Publishing House. New Delhi, pp. 264-272.

Paliwal, H.V., Bhatnagar, S.N. and Haldar, S.K. (1986). Lead-zinc resource prediction in India: An application of Zipf's Law. *Mathematical Geology*, **18(6):** 539-549.

Sarkar, S.N. and Rai, K.L. (2002). Geochemical modelling of sulphide mineralisation in Mosaboni-Rakha Sector of Singhum Copper Belt (Jharkhand) with reference to ore-genesis and mine exploration. *In:* Computer Applications in Mineral Development and Water Resources Management (Eds. K.L. Rai, G.R. Sahu and P. Diwan). South Asian Association of Economic Geologists, pp. 79-93.

Zipf, G.P. (1949). Human behavior and the principle of least efforts. Hafner Publishing Co. New York, p. 573.

Acknowledgement

Compilation of this monogram has been made possible through the assistance and co-operation from a number of organizations and individuals. Grant of Dept. of Science and Technology's scheme on Utilization of Scientific Expertise of Retired Scientist (USERS) to the author over a period of two years from 2002 to 2004 with one Research Assistant enabled him to devote full time to this project work. The Principal, Presidency College, Kolkata and Head of the Department of Geology provided all laboratory and office facilities during the compilation of this monogram.

As regards individuals, I must express my gratitude to Late Professor Ajit Kumar Saha who initiated me for developing the methodology of modelling of mineral deposits during the mid 80s of last century. I must acknowledge the fruitful discussion held with my friends and colleagues while carrying out R&D work on exploration geochemistry during my service period in Geological Survey of India (GSI). Thanks are also due to the faculty members of Geology Department, Presidency College, Kolkata, who directly or indirectly helped me a lot while preparing this monogram. Special mention should be made about Dr. Swapan Haldar, former Chief Geologist, HZL and Scientist, Department of Geology, Presidency College, Kolkata who has kindly discussed with me about exploration of concealed mineral deposits and given valuable suggestions. Shri Partha Pratim Banerjee and other Research Assistants of the Department actively assisted the author while preparing the draft write up.

Some of the figures, tables etc. have been reproduced from various books and journals. Thanks are due to the authors and publishers for kindly permitting me to do so.

Contents

Chapter 1
MODELLING WITH EARTH SCIENCE DATA

1.1 INTRODUCTION

In a country like India various types of earth science data have been generated for about 170 years by the Geological Survey of India (GSI) and subsequently by its much younger sister organizations like Oil and Natural Gas Corporation, Indian Bureau of Mines, Atomic Minerals Division, Central Ground Water Board etc. and a number of public sector undertakings like Hindustan Copper Limited, Hindustan Zinc Limited etc. In addition to these various earth science related organizations, State Departments of Mines and Geology and a number of academic institutions and Geology Departments under the various universities and Indian Institute of Technologies have come up after the independence of India. Hence, over the years, voluminous data have been generated on different aspects of earth science. For example huge amount of data has been generated by detailed petro-chemical studies of granite and associated mineralization and similar other information in different parts of India since middle of twentieth century (Banerji and Talapatra, 1966; Talapatra, 1966; Talapatra, 1968). This is the case for almost all the countries of the world where geological investigations have been taken up. These data are mostly recorded in publications, research notes and unpublished reports of various organizations in different countries. But for meaningful study and utilization, such large volume of data should be documented in a structured form in computer compatible format so that data retrieval from the different modules of the database may be quick and easy. Obviously, this type of database will be a very large one (Fig. 1.1, modified after Talapatra, 2001) with provisions to keep the different types of geological information in a structured form. Then, any type of statistical and geostatistical

© Capital Publishing Company, New Delhi, India 2020
A. K. Talapatra, *Geochemical Exploration and Modelling of Concealed Mineral Deposits*, https://doi.org/10.1007/978-3-030-48756-0_1

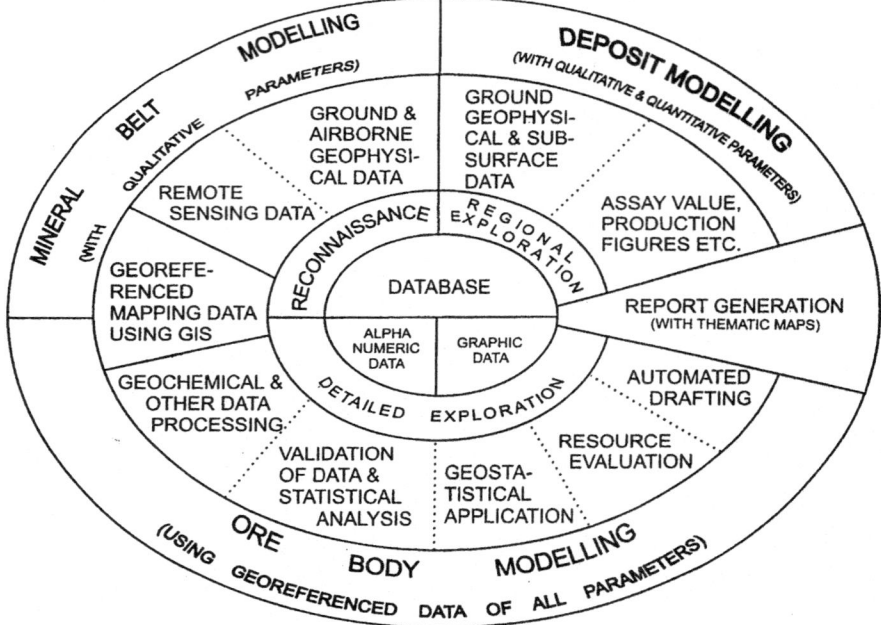

Fig. 1.1 Hollistic diagram showing different modes of data-capture for different stages of mineral exploration

study may be done with these data to arrive at the desired inferences (Agterberg, 1974; Agterberg et al., 1972). This will be further useful if we can carry out computer-based 3-D modelling of mineral occurrence and deposit, which will be discussed in the subsequent chapters.

In fact most of the countries of the world, smaller or bigger one in size, do contain some amount of *in situ* rocks and the major part is generally concealed, covered by soil, alluvium or sand. Naturally the concealed mineral deposits are likely to be present in those covered portions. Now that various Geographic Information Systems (GIS) are available, geo-referencing of the map under study should be done first and all the data stored in the database should have their location co-ordinates (latitudes, longitudes and elevation from the mean sea level) for accurate plotting. Nowadays, field mapping in the land areas for locating an outcrop within the Survey of India Toposheets has become easier due to the availability of hand-held Global Positioning Systems (GPS), which give latitude, and longitude of the location instantly. Similarly, offshore samples are also plotted with the built-in GPS of the Research Vessels along the continental slope or further away (Fig. 1.2). Thus it is easier nowadays to pin-point the field location in the map of any country, whether it is in land areas of any country or along the coastal areas of it.

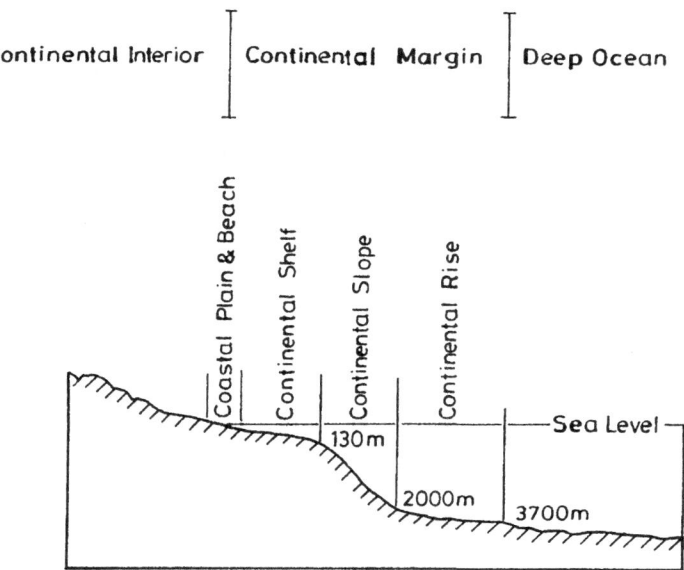

Fig. 1.2   Cross section showing sea beach to continental slope

1.2   WHAT IS MODELLING?

Data generated during systematic geochemical exploration along the terrestrial areas of the world require detailed studies for proper investigation of the related mineral occurrences. In this regard, modelling is a representation of a body in three dimensions, which in the present context is a mineral deposit. The main objective of modelling in this case is to delineate the subsurface configuration of the mineral deposit whether partly or wholly concealed, or in other words the body geometry of the deposit with the help of surface and subsurface data available from the area, especially, those deposits which normally lie hidden below the ground surface. This essentially requires a number of multi-disciplinary types of variable data like geographical, geological, geochemical etc. to build up a structured database so that any type of retrieval can be done for different types of analyses.

Normally, most of the mineral deposits of any country remain concealed below the surface, rarely keeping some surface indications of mineralizations within *in situ* rock outcrops. These indicators are generally weathered, eroded and covered by soil, alluvium, desertic sand etc. mentioned earlier. In order to locate the hidden deposits we require different types of surface as well as subsurface data from the database, along with a geological map at least on 1:50,000 scale. With the help of these data, whether qualitative or quantitative, "Mineral Deposit Modelling" may be build up giving rise to a geological model based on various earth science variables/parameters of a region/belt or a metallogenic province. Talapatra (2001) have discussed in detail how "mineral belt modelling" with the help of qualitative data can be done to locate new target areas in any part of the land irrespective of the country concerned. Then, with the help of additional subsurface quantitative data "deposit modelling" and finally "ore-body modelling" can be done step-by-step with various types of data giving a holistic approach for mineral deposit modelling (Fig. 1.1). These will be elaborated in the subsequent chapters with the case histories where ore body or mineralized zone has been well defined.

A mathematical model taking into consideration various geoscientific data of different inter-disciplinary inputs related to any known deposits follows such geological modelling. Subsequently, this combined model is applicable for estimation of resource potential of the entire area. This will eventually help in predicting the new target areas also in that region/belt or province for further detailed exploration of mineral resources, whether along the continental crusts or along coastal or offshore areas of any country of the world.

1.3 SOME BASIC CONCEPTS

We find different types of models by different authors in the geological literature like conceptual/genetic or empirical model, where the mode of occurrence and origin or genesis of a deposit of a particular mineral belt are described only to predict the likely presence of some mineral deposit and its extension in a qualitative way. Generally this type of model does not predict the probable resource estimates. But in the case of statistical/geostatistical or mathematical model including both probabilistic or deterministic approach, it gives some approximate resource position of the area studied along the mineralized belt concerned with some confidence, provided sufficient surface and subsurface quantitative data are available. Based on these estimates exploiting agencies can start their work for detailed exploration.

A systematic approach to resource evaluation of a mineral deposit is possible using both classical statistical analyses along with geostatistical modelling techniques, taking into consideration the geological controls of such deposits. Classical statistics assume that the sample values are randomly distributed and are independent of each other. A geological database having different field and laboratory data can be processed by classical statistical analysis (including both univariate and bivariate types) through suitable computer programme (Sarkar et al., 1988; Sarkar, 2005). The analysis generates various statistical parameters, frequency histogram plots, correlation matrix etc. that are ultimately utilized for resource estimation. Geostatistical modelling takes recourse to semi-variogram model fitting using the geological database of a mineral deposit. There are various mathematical models to represent the variances underlying a semi-variogram. One such model is the Spherical Model (Clark, 1979) commonly used in mineral deposits, where the sample values become independent once a distance of influence is reached. A procedure for checking the validity of a semi-variogram model that represents the true value underlying experimental semi-variogram of a particular block of the deposit is Point Kriging Cross Validation (PKCV) technique (Davis and Borgman, 1979). In the case of bauxite, iron ore etc. such studies have been done successfully (Biswas, 1997; Kumari, 1996) with Indian deposits.

Geostatistical modelling generally reveals an accurate regional picture of a deposit where sufficient surface and sub-surface data with their location co-ordinates at regular intervals are available. This helps to estimate the extent of a land-based mineral deposit within a mineral belt both laterally and vertically from the available data. Using suitable graphic software package, 3-D body geometry of a deposit can be generated from subsurface data for eventual resource estimation.

Placers may be defined as surficial mineral deposits formed as a result of prolonged mechanical concentration of mineral particles derived from weathered rock mass. The mechanical agent in this process is usually alluvial, but may also be marine, aeolian, lacustrine and glacial in nature (Roonwall 1986). The concentrated mineral is usually of high specific gravity, and it is transported and deposited after the release from the source rock. The sediments that are being carried by river systems from the continental areas generally contain both light and heavy minerals. But heavy minerals are more concentrated along the submerged extension of stream channels or close to the confluence of the river with sea, while the light and less resistant minerals are easily disintegrated and carried away to a greater distance.

Most of the common minerals found in beach placers range in specific gravity from 3.5 to 5.3, excepting a few very heavy minerals like cassiterit (6.8-7.1). These are rutile, ilmenite, magnetite, zircon, monazite, garnet, sillimanite, barite etc. which generally contain significant amount of Rare Earth and Rare Metals along Indian coastal areas. The active wave action of seas generally extends only upto a few metres of water depth in the coastal areas. That is why the placer deposits generally occur along with present day beaches as well as in offshore areas close to the shore, near the confluence of river in different countries of the world. At times, heavy mineral concentrates may be found much further offshore due to past sea level changes, tidal currents, wave actions etc. (Kunzendorf, 1986). To and fro wave action of ripples plays an important part in concentrating and accumulating the heavy minerals along the coastal sandy tracts and shelf areas. Wind movements may also sometimes produce ripples near the coast resulting into placer deposits as in the case of tin deposits in the Malaysian Peninsula where cassiterite rich placer was formed by the disintegration of the host rocks of tin deposits.

It has been observed that most of the world's marine placers were formed during the Pleistocene ice age when the sea level was as much as 160 m lower than the present level (Donn et al., 1962). A series of beaches and extensive channels were developed during this time along the coastal zones in different parts of the earth, because of the cyclic nature of the sea level changes during glacial and non-glacial periods.

In the case of offshore heavy mineral deposits, systematic vibrocore data in a block collected from a vessel can be used for estimating the total resource of the heavy mineral concerned in that block up to certain depth from the sea bed (Fig. 1.3). Generally, vibrocore data of heavy mineral placers are available along the sea coast upto a depth of 4-5 m only. As such heavy mineral resources of the offshore areas can be calculated upto that depth only, close to the territorial water (Fig. 1.4).

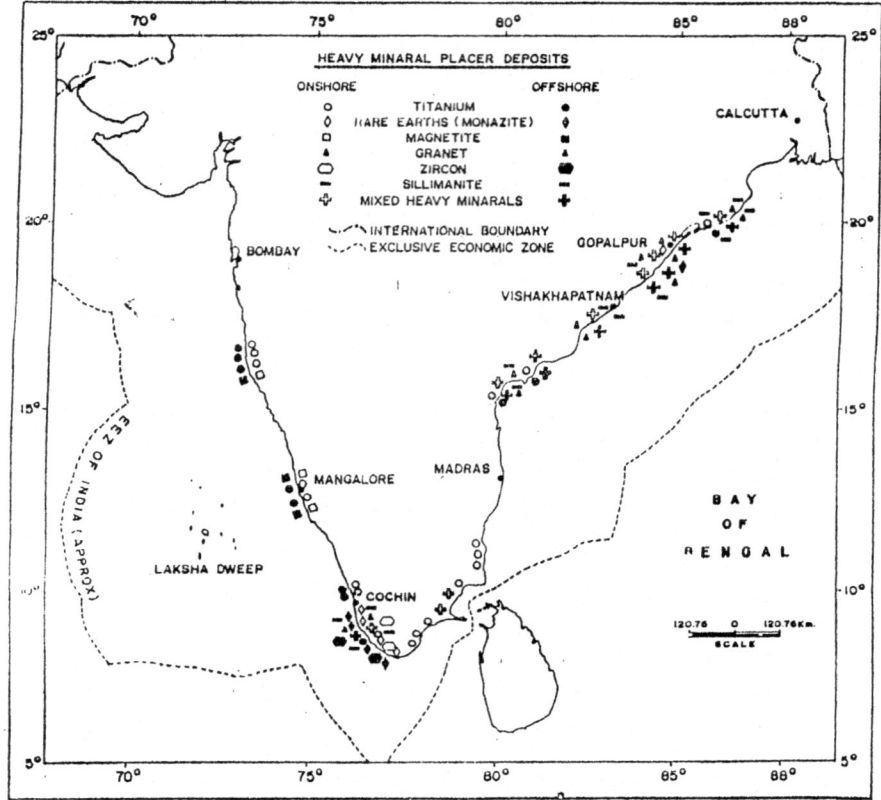

Fig. 1.3 Outline of EEZ of India showing different types of important heavy mineral deposits along the coast

Each core sample of individual block (Fig. 1.5 and Fig. 2.4 of Chapter 2) is generally splitted into two halves of 50 cm long sub-samples, and the bulk heavy mineral percentage of each sub-sample is determined in the laboratory by grain counting method under microscope (Talapatra, 1999), which gives a 3-D picture of the total heavy mineral concentration of the sub-samples of a particular block exaggerating the vertical axis (Fig. 1.5 and Plate 1).

In order to utilize the large amount of data generated by vibrocore and grab samples on Indian offshore areas by Marine Wing of GSI (Sengupta et al., 1992), this 3-D modelling technique was attempted to visualize the body geometry of the deposit (Talapatra, 1999) from the sub-surface data of Block-3 off Gopalpur (Fig. 1.5) which has been described in details in Chapter 2 under "Test Application

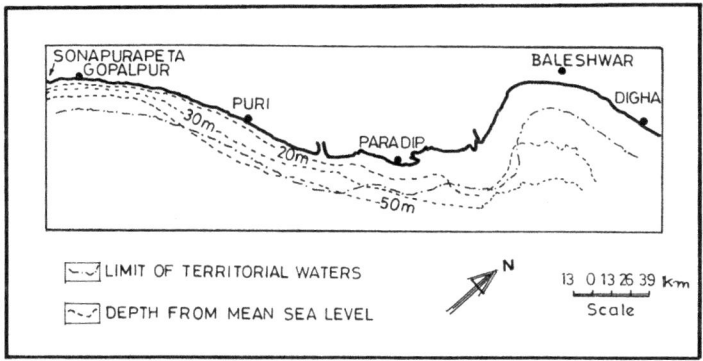

Fig. 1.4 A map showing territorial water zone along the east coast of India between Digha and Sonapurapeta

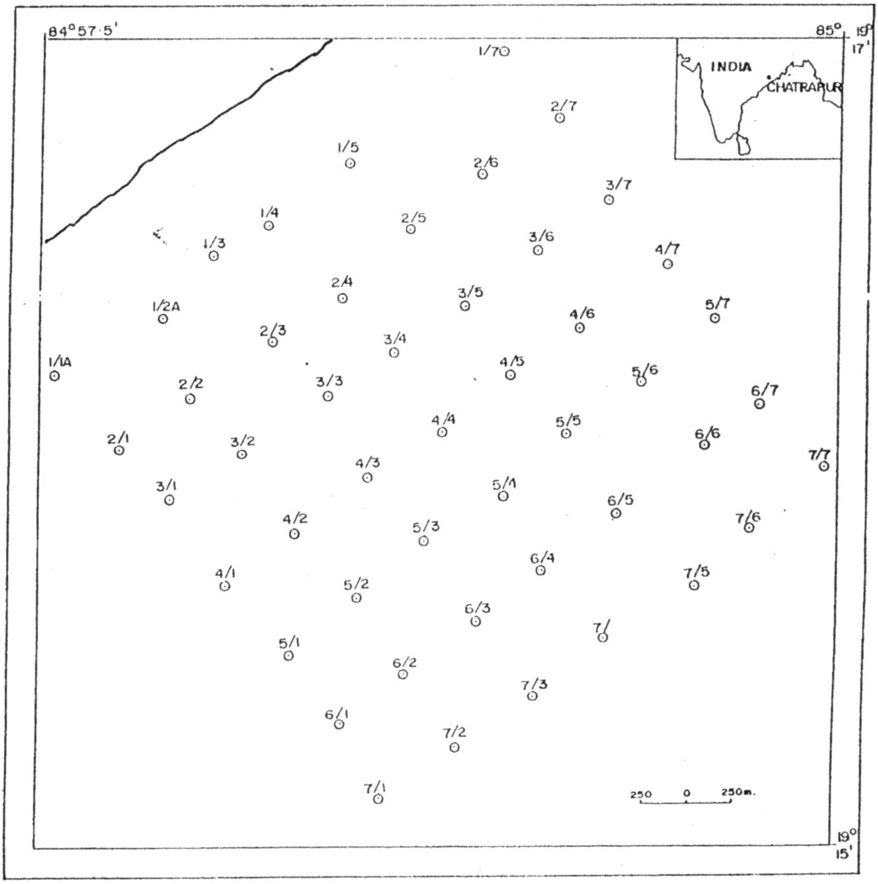

84°57·5'   85° 19'
                                                                  17'

1/70

INDIA
CHATRAPUR

2/7

1/5   2/6
                                        3/7

1/4   2/5
1/3   3/6   4/7

2·4   3/5
1/2A   4/6   5/7
2/3   3/4
1/1A   4/5   5/6   6/7
2/2   3/3
       4/4   5/5   6/6
2/1   3/2   7/7
3/1   4/3   5/4   6/5   7/6
4/2   5/3   6/4   7/5
4/1   5/2   6/3   7/1
5/1   6/2   7/3
6/1   7/2
7/1

250   0   250m.

19°
15'

Fig. 1.5 Location of vibrocore samples and coastline of Block 3 off Chatrapur of Gopalpur area

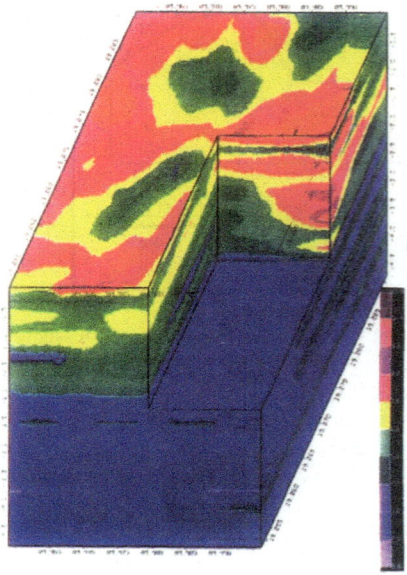
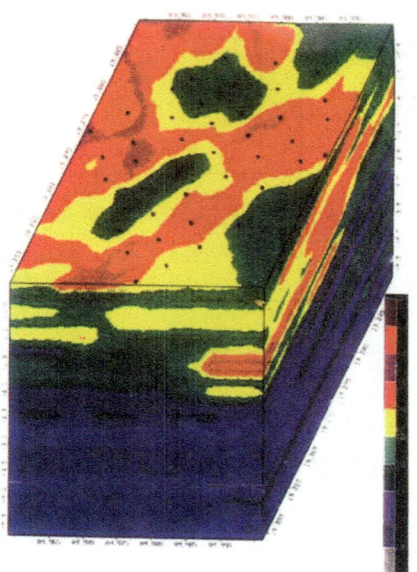
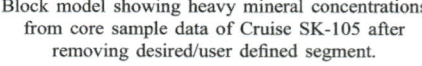

Block model showing heavy mineral concentrations
from core sample data of Cruise SK-105 after
removing desired/user defined segment.

3D block model showing heavy mineral
concentrations from core sample data of
SK-105 Cruise off Gopalpur, Orissa.

Plate 1   3-D colour coded diagrams of Block-3 off Gopalpur showing distribution of total Heavy
Mineral concentration

of 3-D Modelling". Plate 1 shows the 3-D diagram generated to visualize the
body geometry of total heavy mineral weight percentage with depth as well as its
lateral extension from Block 3 off Chatrapur (vide Fig. 2.4) considering the core
samples collected by SK-105 Cruise. The other block diagram in Plate 1 shows
the heavy mineral concentrations after removing desired/user defined segment
from the Block.

Chapter 2
MODELS FOR DIFFERENT TYPES OF DEPOSIT

2.1 INTRODUCTION

Considering the geological setup of India, it is most interesting to highlight the different types of mineral occurrences and deposits present here. Indian sub-continent is blessed with rocks and minerals right from the Archaean to Recent ages. Naturally it contains all the three main types of rocks, namely, sedimentary, igneous and metamorphic variants that include almost all sorts of economic minerals including Rare Earth Elements (REE), Rare Metals (RM), fuels, placer minerals, gemstones and decorative stones in abundance. As such, exploration of different commodities require different types of modelling approach. However, a generalized outline of computer-based 3-D modelling for land-based base metal sulphide mineral deposits will be presented first with some Indian case histories and subsequently modelling of some offshore mineral deposits, mainly heavy mineral placer deposits, that are present all along the vast coastal belt of India will be discussed. Of course, one can easily utilise the procedure for 3-D modelling of any deposit of terrestrial and offshore areas in any part of the world, including mineral deposits of different sedimentary basins, provided sufficient surface and sub-surface data are available for the deposit. In this respect, deposits of south and south-east Asian countries will be highly benefitted by applying these techniques for exploration of essential ore minerals. Before coming to the modelling part, let us discuss in short the important land deposits of the earth.

Various Types of Land Deposits
Depending upon the various types of mineral deposits present in any country, specific strategy for exploration should be designed. Formulation of proper strategy for exploration of land-based mineral deposits requires clear understanding of the

© Capital Publishing Company, New Delhi, India 2020
A. K. Talapatra, *Geochemical Exploration and Modelling of Concealed Mineral Deposits*, https://doi.org/10.1007/978-3-030-48756-0_2

geology of the mineral belt along with its structures and tectonics. In this respect, conceptual/genetic models are quite useful for identifying the target area for outlining the exploration programme. The task has a very wide scope and may be taken up in phases, like (i) Base and precious metals, (ii) Other non-ferrous metals and (iii) Fertiliser minerals etc. The phase I may include study of Precambrian shield areas, younger island arcs and orogenic belts. The Precambrian crustal evolution and metallogeny has been discussed here before going into the details of the important mineral belts of India.

(i) *Concept of Crustal Evolution and Metallogeny of Earth*

It may be mentioned here that over the past 2-3 decades there has been intensive research by the earth scientists on ore locations vis-à-vis tectonic settings, specially after global recognition of the plate-tectonic hypothesis as the most valid concept of crustal evolution. Studies, specially those pertaining to massive sulphide, porphyry copper and Kuroco type deposits in Cenozoic mobile beds, met with significant success and these have led to the framework of concepts relating to metallogeny and crustal evolution in time and space. At the same time, available records contain many instances of spectacular ore discoveries without the aid of any concept. Tracing the source of mineral boulders discovered Athabasca uranium. The Navan Zinc deposit of Ireland was struck following a routine geochemical survey. Stream sediment surveys led to the Red Dog Zinc deposit of Alaska. The geophysical anomalies helped the discovery of Neves Corvo Copper deposits of Portugal. These are the examples of some such deposits.

On the other hand, concepts relating to metallogeny have also led to spectacular and exciting discoveries (Smith et al., 1980). Olympic dam deposit is one such case where the deposit turned out to be very different from what was expected before conceptual reorientation. It inspired confidence during exploration. La Esconda Copper deposits of Chile (a blind ore body) was similarly discovered on the basis of three possible models. The case history of Hemo Gold deposit of Ontario is highly instructive as the association of strata bound gold with volcanic tuff provided the break-through when cores and logs were subsequently reexamined and exploration resumed.

The current thinking about mineralisation is still having opposing ideas. One group is convinced about the high degree of predictability in ore distribution on sound concepts. They see direct and causal links between metallogeny and plate tectonics, and according to them no well conceived exploration programme should fail. The other group holds on the concept "Ore is where you find it" and nature appears to be fond of confounding academic predictions. Despite this scenario it has to be admitted that good geological thinking would be needed for locating concealed ore bodies by new techniques of exploration geochemistry. This will justify capital investment in R&D in this cost spiraling economy.

The developing trends of comprehensive conceptual modelling relating to metallogeny contain the potentials of providing guidelines for regional mineral exploration. In the Archaeans (3800-2500 Ma), two types of terrains are observed: (a) gneiss dominated belts metamorphosed largely to high metamorphic grade (Granite

Gneiss belts) and (b) well preserved, low grade volcanic dominated Greenstone belts. In the former, the most common rocks are granulite to upper amphibolite facies gneisses with remnants of some of the earliest volcanics, layered igneous complexes, anorthosites etc. The main rock units are quartzo-feldspathic gneisses with enclaves of amphibolites and supra-crustals like mica schists, marbles, quartzites and iron formation. The layered complexes are generally of anorthosite, leuco-gabbro or mafic-ultramafic composition. Post-tectonic end-Archaen K-Granites are common.

The greenstone represents the oldest well preserved volcano-sedimentary basin bordered and intruded by granites/gneisses. These may have a linear plan and a basin shaped cross section infolded/downfolded forms in a sea of granitic material. Sometimes, the stratigraphic order is not easy to decipher due to large scale thrusting and nappes. Late Archaean greenstone belts present more complete sequences, less affected by granitisation and later tectonic plutonism. These belts are generally a lower volcanic and upper sedimentary group. In the volcanic group the lower part is ultramafic to mafic components with minor felsic volcanics. The overall composition is thus bimodal. The sediments associated with volcanics are mainly chert, jespar and banded iron formation. The sedimentary group is dominantly clastic with a lower argillaceous deeper water facies (shale, sandstone, grewacke) and an upper arenaceous shallow water facies (conglomerate, quartzite, limestone and banded iron formation). The latter generally occupies the tops of individual cycles. The sediments of older Greenstone belts (Early Archaean) contain silicified volcaniclastic, banded carbonaceous cherts, silicified evaporites and stromatolites; and the volcanic sequences Algoma-type iron formation, hyaloclastic breccias and massive sulphide deposits. The late Archaean Greenstone belts, however, have more deep water facies and terrigenous clastic intercalations. Algoma-type iron formations and massive sulphides are common to both. In the volcanic sequences carbonaceous cherts, evaporites and stromatolites are rare. Differences between the two types reflect increasing influence of sialic crust with time.

Various models have been suggested to explain the evolution of the Granulite-high grade gneiss and Greenstone-Granite belts. The evolution of Greenstone-Granite belts has received more attention. Many variants of modern style plate tectonic processes have been conceived to explain the features and evolution of Greenstone belts by a large group of workers. In the older Greenstone belts (which are deformed and only fragmentary relics are preserved in a sea of gneisses and granite diapers), no basement can be identified. The predominance of mafic and ultramafic rocks in the sequence indeed suggests those to be early oceanic crust but sialic components are invariably present in the neighbourhood and often reveal traces of Pre-Greenstone deformation. Thus the ocean basins were either small or floored by the older sialic material. The younger Greenstone belts are best explained by the plume model of Lambert and Groves (1981) and rift and sag model of Goodwin (1977, 1981), which was designated as Archaean plate tectonic model. Rifting would cause subsidence of the basin. This process is enhanced by downward drag of the under-plated mafic-ultramafic volcanics. This sagging process eventually leads to decoupling with further production of basaltic magma. At that stage fusion and crustal contamination lead to andesite, highly differentiated volcanics as well as consanguous calc-alkaline granitoid intrusives. If crust thins out sufficiently due to stretching and sagging it may break apart giving rise to ensigmatic greenstone belts (Goodwin, op. cit.). Fissuring and rupture may resemble modern rifts generated small ocean basin of modern Red Sea type. Even in such cases of complete plate separation and ensigmatic Greenstone belt generation the high oceanic geotherms at that time must have kept the crust buoyant.

Analyzing the various suggestions and evidences Kroner (1981) concluded that the high grade terrains represent gneissic sialic crust with predominantly sedimentary deposits and granitoids produced during incomplete Pre-Greenstone fissuring and subsidence. Greenstone evolution (over plating) and/or crustal underthrusting/ inter-stacking moved much of the material into lower crustal levels with transformation to migmatites, banded gneiss, granulites and grey tonalite gneiss. This type of lower crust underlies most of the granite-greenstone terrains, or may have been brought in juxtaposition by thrusting etc. Base and precious metal ore deposits are comparatively rare in granulite-gneiss belts. Minor copper occurrences have been reported associated with the relict banded iron formation and amphibolite/granulite facies. The amphibolite enclaves in granulite-gneiss terrains do rarely contain some Ni-Cu deposits. The best example is that of Selibi-Pikwe (Limpopo belt), where the mineralisation is evidenced by rusty sulphide zones in amphibolites. Reconnaissance of amphibolite enclaves in granulite-gneiss terrains may thus be worthwhile.

Greenstone belts are known to be repositories of Au, Ag, Cr, Ni, Cu and Zn ores and these metallogenic events can be more systematically related to tectonic evolution. The ore deposits occurring in Greenstone Granite belts are remarkably similar in the different continents. Windley (1984) observed that the mafic to felsic volcanic sequences of the Greenstone belts have been responsible for the major Cu, Zn, Au and Ag deposits of the world. The volcanic exhalative ores are stratigraphically controlled but their final locations are structurally controlled. The ultimate source of Au, Cu and Zn is in the volcanic pile (mafic-ultramafic), but ore localisation is due to mobilisation by thermal gradients set up by the crustal granites. Gold and silver are traditionally related to regional fractures connected with stocks, lava flows and unconformities. Gold in banded iron formation is located to aguagene tuffs, and is precipitated from sub-aqueous volcanic exhalations. Most important gold deposits occur in the marginal zones of Greenstone belts near granitic plutons, and fade away from such zones. Silver commonly occurs with gold (e.g. Abitibi – Ag with Au, Cu, Zn) at felsic-mafic volcanic contacts. Some of the granites do not contain porphyry type copper-molybdenum deposits. Copper and zinc sulphides with occasional lead association generally occur in the volcanics and pyroclastics in the upper parts of the volcanic cycles (e.g. Abitibi belt).

The Proterozoic (2500-600 Ma) sequences of the earth have also been analysed from the angle of crustal evolution. In this respect the contributions of Kroner (1981) and Windley (1984) seem to be quite comprehensive. Kroner (1981) identifies a two-fold division: (a) Lower Proterozoic (2500-1200 Ma) and (b) Upper Proterozoic (1200-600 Ma). Windley (1984) lists the following stages of Proterozoic developments: (a) Early to Middle Proterozoic basic-ultramafic intrusions, (b) Early to Middle Proterozoic basins and belts, (c) Mid-Proterozoic anorogenic magmatism and abortive rifting, (d) Middle-Late Proterozoic basins and dykes, and (e) Late Proterozoic Mobile belts. According to Kroner (1981) the data on the time interval 2.5-2.2 Ga (i.e. lower Proterozoic time) do not reveal significant world-wide tectonic activity. This period was characterised by development of large sedimentary basins on stable continental crusts. But Windley (1984) identified early Proterozoic rifting characterised by Great Dyke like layered bodies which intruded into the consolidated Greenstone belts and high level granites. He also reported swarms of dolerite dykes cutting across Archaean high grade rocks and Greenstone belts, and layered stratiform complexes (like Stillwater complex, Sudbury complex, Bushveld complex) which intruded the older basement and Early Proterozoic cover. According to him the formation of thick continental crust at the end of the Archaean made it possible for the first continental margin marine basins to develop. In these new tectonic environments six principal types of sedimentary ore deposits formed are: (1) Au-V in conglomerate, (2) Mn and Pb-Zn in carbonates, (3) Banded iron formation (Superior Type), (4) Cu-U-V in clastics, (5) Evaporites and (6) Phosphorites; of particular interest is the Pb-Zn hosted dolomites typically belonging to a shallow marine intertidal lagoonal to littoral facies often with algal reefs that is formed in a carbonate shelf-setting at rifted continental margin. Such dolomites provide the host for Cu-Pb-Zn mineralisation at McArthur River and Mt Isa type of deposits. The rift related Late Proterozoic mineralisation of stratiform Cu in Zambia, Cu-breccia types in Trilog, Ontario and magmatic Cu-Ni in Duluth Complex are considered to be of similar type. In addition Windlay (1984) identified Cr-Ni-Pt-Cu mineralisation associated with the basic and ultrabasic rocks of early Proterzoic intrusions and Cu- mineralisation in basaltic lavas related to Mid-Proterozoic anorogenic magmatism.

Plate movements are active from Precambrian to Quaternary time and it has been demonstrated that crustal movement due to plate-tectonics have clear bearing on metallogeny and as well as neo-tectonism giving rise to recent earthquakes. Hence plate-tectonic rifting, continental separation, ocean floor generation and subsequent subduction, collision and closure of plates are very important from the point of view of exploration of mineral deposits along the related metallogenic belts.

(ii) *Potential Mineral Belts of India in Geotectonic Frame*

Going through the various concepts of crustal evolution and metallogeny described above, it is now necessary for arriving at a logical basis for delineating potential mineral belts of India which aims at identification of targets for base and precious metal exploration with special reference to Precambrian shield. Within the Achaean terrains of India granulite-gneiss and greenstone granite belts can be identified. Rift and sag or sag-subduction was the dominant process of Greenstone belt evolution, Ganulite Gneiss belts were broadly coeval attaining the high grade of metamorphism in lower crust but now juxtaposed with the Greenstone belts due to thrusting/unroofing. It is also necessary to recognize that the marginal basin model of Greenstone belt evolution is highly probable.

Granulite-Gneiss belts of India are so far found to be metallogenetically poor excepting that amphibolite enclaves at places contain Ni-Cu sulphide deposits. Some of the supracrustal meta sediments of this lower crust might have been raised to granulite facies condition and are, thus, now inseparable from their basement rocks. These metasedimentary segments, distinct from the Greenstone, may contain base metal suphide deposits. On the other hand Greenstone-Granite belts are normally repositories of Au, Ag, Cu and Zn ore deposits. The source of these metals is the volcanic pile of Greenstone belts, subsequently mobilized to suitable locale by thermal gradients and deformation and structural configuration. Quartz pebble Au-U deposits are well known from the base of Greenstone lavas which may be present in Indian condition. Volcanogenic stratabound massive sulphides of Cu, Zn, Au and Ag associated with the rift related felsic volcanics and sulphide and carbonate facies iron formation are also expected.

The dominant process of Proterozoic crustal evolution was ensialic orogeny. Lower-Middle Proterozoic encrustal evolution was dominated by rifting, development of passive continental margin, failed rifts/aulacogens and limited ocean opening and crustal shortening which can be delineated in Indian tectonic set up. Metallogenetically the Proterozoic is most marked by sedimentary exhalative (SEDEX) and volcanogenic type ore deposits, typically related to specific paleotectonic environments. Exalative sedimentary (shale hosted or carbonate hosted) Pb-Zn-minor Cu sulphide deposits also occur in shallow marine inter-tidal lagoonal to littoral sediments. The mafic ultramafic complexes resulting from anorogenic magmatism during early Proterozoic carry important deposits of Cr-Ni-Pt-Cu and the mid-Proterozoic carbonatite hosting P, Nb, U and Cu mineralisation.

The Indian shield has cratonised primarily through Archaean and end- Archaean crustal evolution processes. Some areas of the shield, however progressed, through rifting to fold belt generation in the Early-Middle Proterozoic that led to accretion and welding of Archaean crustal segments. In Late-Proterozoic extensive shelf and platform sequences developed on the craton.

This geotectonic development can be traced in tectono-stratigraphic record of the individual mobile belts and their nucleus area of India. On a regional scale, however, four segments can be recognized in Indian Shield that contain characteristic evolutionary history. These are the South Indian shield, the Central Indian, the Western and North Indian shield. The south, central and eastern segments are separated from the north and west segments by the most prominent Narmada-Son lineament, which is a very significant tectonic feature. The eastern, central and the southern segments are again tectonically separated by the Mahanadi and Godavary-Gondwana grabens, respectively. However, subsequent Proterozoic reactivation zones may transgress the tectonic boundary between the segments. Potential mineral belts of these four segments of the Indian shield will be described in short in the following paragraphs.

The South Indian Shield is consisting of three principal components, namely the Archaean high grade gneisses with reworked variants, the Archaean green stone supra crustal and the Proterozoic Cuddaph cover sequence. The South Indian craton evolved through Archaean accretionery processes with formation of granite greenstone terrains and the lower crustal granulite facies domains. Base metal mineralisation with nickel in the high grade segments is sporadic in Tamilnadu as in Pattankadu, Tirunellvelly District and Arunmanathur, Kanyakumari district and also in some parts of Madurai district. The only important marginal base metal deposit located so far is at Mamandur, where Zn, Pb and Cu mineralisation occurs at the contact of migmatite and garnet biotite sillimanite gneiss.

The South Indian Archaean granite-greenstone terrain can be again divided into four provinces on tectonic style, stratigraphy and metamorphism. They are (i) the southern high grade Sargur type, (ii) the eastern Kolar type containing granite-greenstone belt, (iii) the western Shimoga Goa type and (iv) the Northeastern Nellore type. The latter three types are important for base metal and gold mineralisation besides their iron and manganese potential. Since crustal development during the Archaean time is responsible for the degree of metal concentration when the upper part of mantle vigorously fractionated. Hence the Greenstones are metallogenically important and they retain the vestiges of early mantle and the obvious targets for copper and gold mineralisation.

The crescent-shaped Cuddapah basin is a middle Proterozoic rift structure containing fairly widespread rift-related igneous rocks represented by basic and felsic volcanics and dykes in the lower Cuddapah sequences of the western and southern parts of the basin. A major thrust belt separates Cuddapahs from the basement along the eastern contact. The Nallamalai sub-basin in the east has evolved into a fold belt due to this deformation, and this fold belt contains the majority of the base metal occurrences associated with a carbonate-chert facies overlain by carbonaceous shale of probable tuffaceous affinity. The Nallamalai lead-zinc deposits appear to be exhalative sedimentary type which has been enriched by mobilisation.

The major part of the Precambrian geology of Central Indian shield is covered by Vindhyans and their equivalents, the Gondwanas and the widespread Deccan Traps. One of the most interesting features of this region is the intersecting mobile belt relationship between the north-south trending Bengpal-Bailadila-Dongargarh tectonic feature with the Sausar-Raigarh east-west Satpura trend giving rise to Bhandara triangle which is occupied by the Sakolis. The significance of this type of mobile belt intersection is not clearly known but the feature seems to be linked with superposition of east-west Early- Middle Proterozoic rift structure of the Sausar-Chilpi belt, its southern arm having faulted on an older north-south trend showing an early Archaean granite greenstone development. Such broad sequence of crustal development is metallogenically significant and points to some indication of probable potential belts for long and short term investigations, namely Bengpal belt, Betul belt, Malanjkhand extension belt, Sakoli-Surjagarh belt, Mahakoshal belt etc.

The Eastern Indian shield comprises an extensive area of reworked Archaean granite-greenstone complex in Chotanagpur and Eastern belts, where remnants of Archaean Greenstone components are recognisable. This belt encloses the Early Archaean sialic nucleus in Singhbhum. A Late Archaean to Early Proterozoic shelf sequence with Banded Iron Formation (BIF), volcanics and Greenstones has developed in the margin of this nucleus in the Gorumahasani and Bonai Groups. The Singhbhum craton is girdled in the north by the Early-Middle Proterozoic mobile belt where a pattern of encialic rifting (Dhanjori-Simplipal) and subsequent convergent tectonics with proto-ophiolite emplacement in the Dalma belt can be recognized. The Kuilapal type granite probably indicates the magmatic response of northerly active Middle-Proterozoic oceanic subduction beneath the Chotanagpur complex. Extensive reworking of the granulite-greenstone terrain of the Chotanagpur plateau also took place during this period. The Singhbhum Shear Zone hosting important Cu and U mineralisation is an ensialic under thrusting (cratonic subduction) zone developing imbricate ductile shears that extensively mobilized and localized the metals from the Dhanjori type ensialic rift related volcanics. These tectonic events can be established in a sequential order provided sufficient geochronological data are available from the chief rock types of the different tectonic units. However, on the basis of geotectonic interpretation several potential mineral belts are identified. Most important among these are Chotanagpur belt, Adash Reactivation belt, Gorumahisani-Badampahar belt, Simplipal-Notapahar belt and Singhbhum Cu belt.

The Western and North Indian shield consist of a number of Archaean and Proterozoic fold belts, which girdle the Bundelkhand massif in the west and the south. These are overlapped by the Vindhyan cover sequences, and the Deccan Traps. The basement of Archaean rocks (3.5 Ga in Rajasthan) have magmatism (Untala and Gingla granites, 2.9 Ga) and End-Archaean K-granite emplacement (Berach granite, 2.5 Ga, and parts of Bundelkhand granite). An Archaean granite-greenstone evolutionary sequence can thus be recognized in the Western and North Indian shield. The proterozoic geologic history of this sequence is full of repeated rifting, at places ensigmatic, with development of diverse stratigraphic and magmatic rocks.

The Early Proterozoic ensialic rift sequences are represented by the Aravallis in Rajasthan and Gujarat, and by the Bijawars in U.P. A characteristic feature of this sequence is the presence of carbonate, BIF, phosphatic beds and Pb-Zn dominated stratigraphy in different parts of the rift basins. The potential metallogenic belts of this shield in the geotectonic frame discussed above have been identified with reported base and noble metal mineralisation. Important mineralised belts are Mangalwar belt (including the biggest Pb-Zn-Cu bearing Agucha deposit, Rajasthan), Hindoli belt, Jahazpur belt, Aravalli Fold belt, Bijawar belt and North & South Delhi belt (including Khetri Cu belt). Potential metallogenic belts mentioned above from the Late Archean to Early Proterozoic age give clue for detection of various mineral deposits in different parts of the world. Naturally, such approach will also be very much helpful for locating even concealed mineral deposits in any part of the earth.

2.2 METHODOLOGY OF MODELLING OF LAND DEPOSITS

Before describing the methodology of modelling let us discuss in details how to search the mineral deposits of any country. Search for mineral deposits is generally taken up after systematic geological mapping (normally 1:50,000 scale) of any terrain provided there is some significant indication of mineral occurrence/concentration in the area. This is followed by large scale mapping of the area of interest after making photogeological studies in conjunction with other available remote sensing data. If there is any positive indication of mineralisation, systematic geochemical sampling, along with ground geophysical studies, is taken up to locate anomalous zones (Fig. 1.1). At this stage pitting, trenching, test drilling and subsequently detail drilling is carried out along the anomalous zones for getting subsurface data, which may indicate the tentative configuration and extent of the causative body of the anomalous zones. Subsequent detailed subsurface probing of these zones may establish the 3-D body geometry of the mineral deposit. For a virgin area, prior study of landsat imagery and airborne geophysical data may be of help in selecting the area of interest before conducting systematic geological investigation. This is followed by detailed geological mapping, accompanied by ground geophysical survey and systematic geochemical sampling. Mineral investigation for locating deposits is generally conducted through exploration methods like pitting, trenching, drilling etc. mentioned earlier.

If along an area with significant lateral extent, numbers of mineral occurrences are reported at certain interval of a few kilometres apart along the strike direction of the litho units, we can call it a mineral belt. Modelling of a mineral belt initially requires qualitative data, which records the presence or absence of particular variables. For this, detailed geological map with all the structural element is essential. If possible geochemical/geophysical data (anomaly etc.) may also be utilized at this stage for differentiating anomalous areas. Any type of modelling requires a multi-modular database of the area of study comprising alphanumeric information along with a digitized and georeferenced map and other graphic data representing different geoscientific parameters stored in computer in a structured form. The suggested scheme of computer-based mineral deposit modelling (Talapatra, 2001) starts with reconnaissance stage information. Here the different variables/parameters from geological mapping, ground and airborne geophysical studies and remote sensing applications are judiciously utilized. To start with, the modelling takes into account only qualitative variables/parameters of the study area like geological mapping data (including presence or absence of litho-units, structures, ore minerals etc.), remotely sensed airborne and ground geophysical information data to build up a database for mineral belt modelling.

In this chapter, base-metal deposits of Pur-Banera-Bhinder belt, Rajasthan has been examined using the methodology already elaborated, and the results have been detailed in the following paragraphs that gives some possible indications of mineralization. Similar treatment of modelling methodology with the bed rock geochemistry of samples collected from northern parts of Purulia-Bankura area, West Bengal has also been described in details in Chapter 6, where minor base-metal deposits along with rare earth and rare metal mineralization are reported. Such studies provide an excellent picture of the ore potentiality of Purulia-Bankura area for further detailed exploration.

2.2.1 *Mineral Belt Modelling*

While describing the methodology of modelling, work conducted by the Field Technique Research Unit (FTRU) of GSI in seventies of last century may be referred for follow up action. Talapatra et al. (1986a) developed a mineral belt modelling based essentially on geological data relating to the Pur-Benera-Bhinder belt of base metal sulphide mineralisation in Bhilwara, Chittorgarh and Raisamand districts of Rajasthan based on qualitative data. An attempt has been made to extrapolate the relationship between known and unknown areal blocks on the basis of "Weighted pair-group average clustering method" (Davis, 1973) using the matrix of matching coefficients, generated from the present/absent data of individual blocks or cells of known mineralized belt (Talapatra et al., 1986a). This method is expected to help in arriving at a statistically "probable" prediction of potential areas (Colliyer and Merriam, 1973). A grid of cells of equal size (here 10 km × 10 km grid) drawn on tracing paper has been superimposed on a geological map along a known mineralized belt (Fig. 2.1). A set of different variables/parameters like lithological, mineralogical, geophysical etc. are used as statistical variables. Then, a present/absent evaluation of these variables for each cell, including the control cells containing proven mineral deposits or known occurrences of one or more commodities, is made so that each parameter of each cell is matched consecutively with the corresponding parameter of all the cells. The method thus makes a comparative study between the geology of the control cells and that of other cells whose ore potential is to be predicted. Accordingly, a unique matrix of matching coefficients was generated as per the following formula:

$$M = \frac{P+N}{P+N+U}$$

where M is matching coefficient, P = number of positively matched cells, N = number of negatively matched cells, and U is number of unmatched cells.

The size of this matrix will be controlled by the number of cells under study and obviously it will be a symmetrical matrix. Once this matrix is generated, cluster analysis may be performed (Davis, 1973) and the results of the analysis may be shown by a "dendrogram" exhibiting interesting clusters of different cells with their similarity levels (Fig. 2.2). The dendrogram showing pair-wise relationship expressed by similarity coefficients is a graphic representation showing ordering between the pairs of cells. The pairs are ordered to bring the most similar next to one another by single weighted pair-group average linking. The purpose of cluster analysis is to show optimum pair-wise arrangements of the interrelationship between the cells within the matrix of matching coefficients. The essential features of this analysis can be summarized as below:

1. Use of the matching coefficients as similarity measures.
2. Determination of similarity levels amongst the different cells.

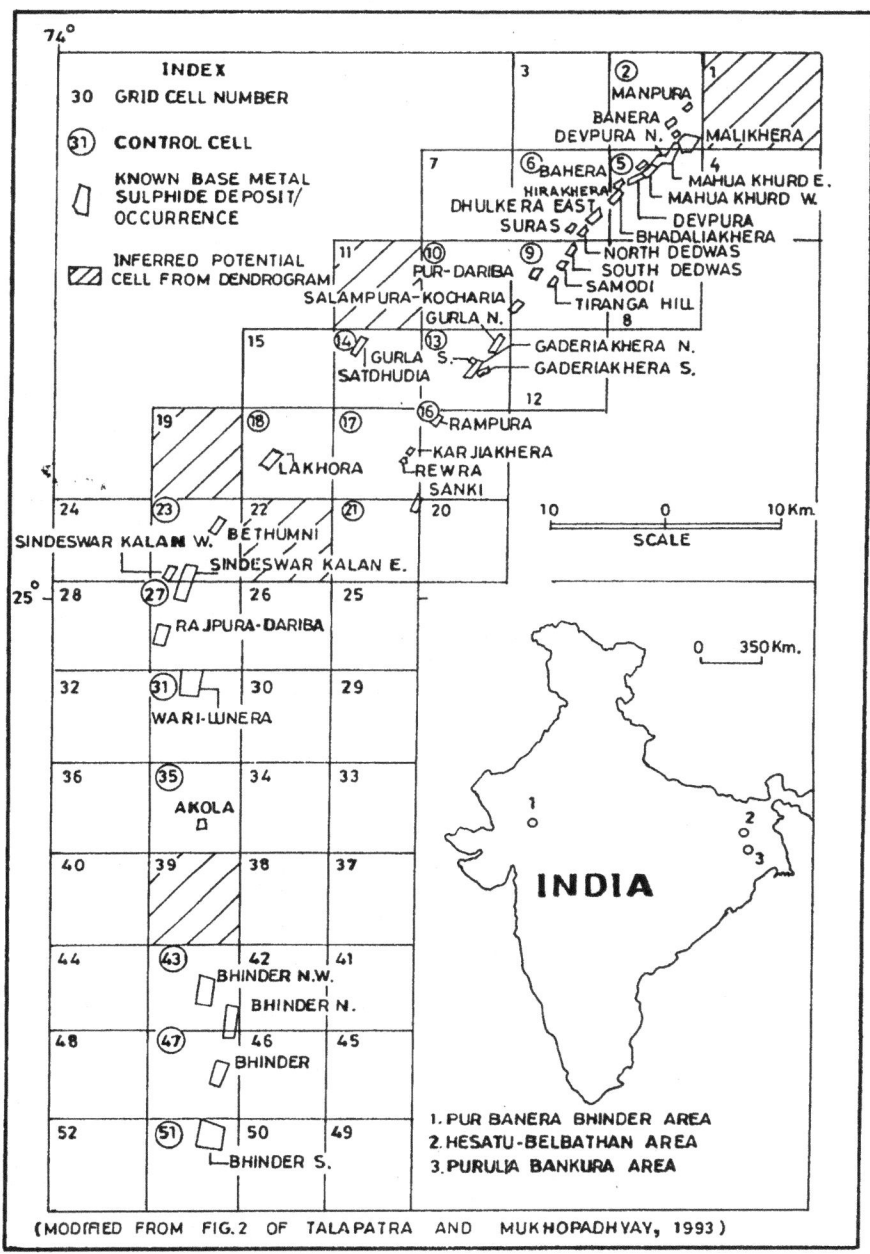

Fig. 2.1 Map showing the different cells along the Pur-Banera-Bhinder belt, Rajasthan with an inset map of India showing three Precambrian mineralized belts studied by the author

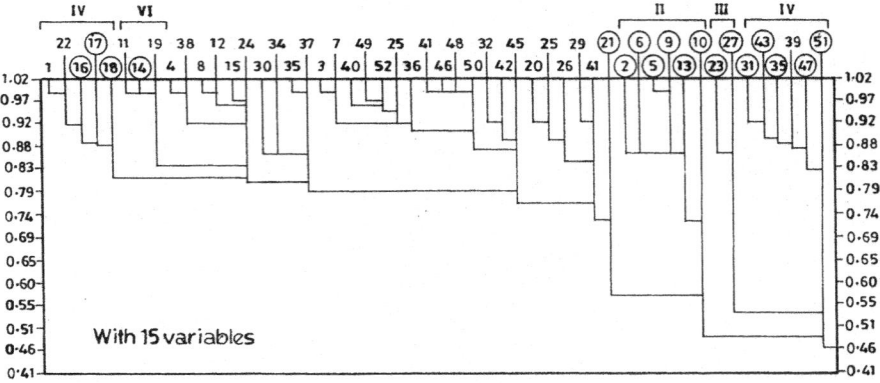

Fig. 2.2 Dendrogram with 15 variables/parameters showing the different clusters

3. Interlinking of two cells at a time, based on mutually highest correlations with each other.
4. After two cells are clustered, their correlations with all other cells are averaged.

To start with, a north-south grid of 10 km × 10 km size cells in tracing paper is superposed on the geological map of the area on 1:253,440 (compiled by Western Region, GSI) and the variables/parameters like lithology, mineralogy, geophysical anomaly etc. have been used. Eighteen control cells containing known deposits/ occurrences of Cu, Cu-Pb and Pb-Zn have been delineated from a total of 52 cells— all in the vicinity of the mineralized belt under consideration (Fig. 2.1). Detailed exploration of the deposits/occurrences of mineralisation within the control cells have been carried out by GSI. At times one control cell may have more than one deposit/occurrence. The variables/parameters pertinent to geological features of the mineral belt are chosen so as to ensure highest representation in each cell parallel to longitude and latitude of the entire mapped area under consideration. Table 2.1 shows the list of variables/parameters.

The present/absent data for each of the variables/parameters are tabulated for all the 52 cells represented by '1' or '0', as the case may be, to form a two dimensional array in which the rows represent the parameters and the columns represent the cells. So, C (I, J) denotes the Ith variable in the Jth cell of the matrix C. Data for each set of parameter is matched against the corresponding data of every cell, and the matching coefficients are computed programmatically as per the formula already mentioned. The matching coefficients will form a similarity matrix of dimension N × N, N being the number of cells. From the similarity matrix, clustering of cells are performed by "weighted pair-group average" method (Davis, 1973) and finally, a dendrogram linking the cells and group of cells at different similarity levels is produced (Fig. 2.2).

In order to examine the influence of the different variables/parameters in the formation of different clusters of cells, the FORTRAN program (Talapatra et al., 1986a) was run, in this case three times with 21, 17 and 15 parameters by three trial

Table 2.1 Different variables/parameters studied along Pur-Banera-Bhinder belt, Rajasthan

Sl. Description No.	Sl. Description No.
1. Granite, granitic gneiss and aplite.	11. Calc-gneiss, calc-silicate, mica schist, marble
2. Staurolite-kyanite mica schist	12. Quartzite, conglomerate
3. Sulphide bearing graphite mica schist and calc-silicate marble with gossan	13. Amphibolite
4. Sulphide bearing mica schist, calc-silicate marble	14. Calc-biotite, actinolite schist, garnetiferous mica schist, ferruginous quartzite
5. Sulphide bearing banded magnetite	15. Dolomitic marble
6. Quartzite	16. Copper ore
7. Mica schist	17. Lead-zinc ore
8. Mica schist, quartzite and marble	18. Lead-zinc-silver ore
9. Carbonaceous mica schist	19. Copper-lead-zinc ore
10. Migmatite, composite gneiss, feldspathised mica schist	20. Electromagnetic anamoly
	21. Strata-bound mineralisation

runs (vide Table 2.2), respectively, thereby omitting some of the parameters. The dendrograms, thus generated, for all the three runs show interesting relationships of different cells (Talapatra et al., 1986a).

The exercise with 15 parameters with omission of parameters listed at Sl. Nos. 6, 7 and 16-19 (vide Table 2.2), appears to be more significant compared with other two and the resulting dendrogram is shown in Fig. 2.2. The following clusters are noted:

1. *Cluster I* consists of cell Nos. 31, 35, 39, 43, 47 and 51 with levels of similarity varying from 0.82 to 0.93. Cell No. 39 is again clustered here.
2. *Cluster II* consists of cell Nos. 2, 5, 6, 9, 10 and 13, which are all control cells. Similarity levels of this cluster vary from 0.73 to 1.00.
3. *Cluster III* consists of cell Nos. 23 and 27 with similarity level at 0.86.
4. *Cluster IV* consists of cell Nos. 1, 16, 17, 18 and 22 with similarity levels varying from 0.88 to 1.00. In this cluster, cell No. 22 seems to be promising because it is adjacent to the control cell No. 17. Cell No. 1 is also significant, as it has clustered with this group.
5. *Cluster V* not recognized as a separate group in this study.
6. *Cluster VI* a small cluster, with cell Nos. 11, 14 and 19 with similarity level at 1.00, is clearly discernible. Because of clustering with control cell No. 14, cell Nos. 11 and 19 may be of some interest from the point of mineralisation.

A synoptic table showing different clusters generated by varying the number of parameters is presented in Table 2.2. It is observed that cell Nos. 23 and 27 containing Dariba-Rajpura (Pb and Zn) have been clustered in all the three exercises and the similarity level increased from 0.81 to 0.86 by reducing the number of parameters from 21 to 15. Similar pattern is observed for cells containing Pur-Banera (Cu),

Table 2.2 Synoptic results of cluster analysis of 52 cells of Pur-Banera-Bhinder belt, Rajasthan

No. of parameters	Base metal belts	Akola-Bhinder belt (Cu)	Pur-Banera belt (Cu)	Dariba-Rajpura belt (Pb-Zn)	Rampura-(Rewara)-Lakhora belt (Zn-Pb)	Salampura-Kocharia (Pb-Zn)	Satdhudia (Cu)
21		31, 35, 39, 43, 47	2, 5, 6, 9, 13	23, 27	16, 17, 18, 21	10, 15	-
		(0.74-0.86)	(0.83-0.95)	(0.81)	(0.72-0.90)	(0.90)	
17		31, 35, 39, 43, 47, 51	2, 5, 6, 9, 13	23, 27	16, 17, 18, 21	-	11, 14, 19
		(0.62-0.94)	(0.84-0.94)	(0.88)	(0.75-0.88)		(0.94-1.00)
15		31, 35, 39, 43, 47, 51	2, 5, 6, 9, 10, 13	23, 27	1, 16, 17, 18, 22	-	11, 14, 19
		(0.82-0.93)	(0.73-1.00)	(0.86)	(0.88-1.00)		(1.00)
Cluster no.		Cluster I	Cluster II	Cluster III	Cluster IV	Cluster V	Cluster VI

N.B.: All cells except those underlined are control cells; similarity levels are given within parenthesis where similarity level and its range have increased from 0.83-0.95 to 0.73-1.00. In the Akola-Bhinder belt also similarity level increased from 0.74 to 0.86 and the higher limit from 0.86 to 0.93. In Rampura-(Rewara)-Lakhora (Zn-Pb) the rise in similarity level is observed (Table 2.2). In addition to these, one small cluster has formed with Satdhudia (Cu) occurrence.

It may be mentioned here that size and orientation of the cells may be changed by trial and error method and then the same exercise may be repeated to see that whether any better clustering of cells are formed for selecting any barren cell for detailed exploration.

Adopting similar procedure, data related to two other Precambrian belts namely, 50 grid cells of Hesatu-Belbathan belt, Jharkhand of eastern India with 20 geological variables (Talapatra et al., 1991) and 49 grid cells of similar size along with 22 geological variables of Purulia-Bankura belt, West Bengal (vide Chapter 6) were treated for cluster analysis and dendrograms were generated as before. These two Precambrian mineralized belts of Eastern India also showed interesting results successfully identifying the clusters of control cells containing base metal and other mineralisation and have shown the possibility of locating new sites of mineralisation within the so-called barren cells (cf. Talapatra, 1991).

A few such areas of barren cell from all these studies have actually shown some potential as revealed by subsequent work of the Geological Survey of India (personal communication) and others are yet to be tested. In this connection it may be mentioned here that similar processing of such data may be done by changing the size and orientation of the grids, separately.

From the above study, it may be concluded that despite limitations inherent to choosing of parameters, orientation and size of cells, assumption of continuity in mineralisation etc., the procedure provides an excellent picture by the qualitative aspect of the ore potentialities of the area. The scope for further enhancement of the method by repetitive trial and error method, using various combinations of parameters by changing grid size and orientation etc. is well within possibility.

The exercises carried out relating to similarity grouping as mentioned above led to the following generalization in resources highlights:

1. Generation of clusters of cells showing characteristic ore mineral assemblages e.g. Cu, Cu-Pb-Zn, Zn-Pb etc., corroborated by the different exercises. The cells having similar type of sulphide mineralisation within the belt under investigation have shown separate clusters.
2. Improvement of the grouping of cells with characteristic mineral assemblages and enhancement of the level of similarities of the clusters on elimination of some variables have been established leading to identification of important predictors out of the different variables for mineralized areas which provides scope for similar studies in adjacent areas of this belt or in different other similar belts.
3. The study has brought out possible indications for mineralisation in some cells hitherto described as "barren" within these mineralized belts. This modelling technique is recommended for quick preliminary mineral appraisal of a mineral belt during reconnaissance stage at low cost.

2.2.2 Deposit Modelling

In any country thorough search of mineral deposits is essential at the beginning.
Once a mineral belt is identified with a number of occurrences or deposits of some
mineral commodity, regional exploration should be carried out one by one (vide
Fig. 1.1) taking into account all the qualitative and quantitative variables/parameters
after proper geo-referencing of the study area using a standard GIS package as mentioned earlier. This is followed by large scale mapping of the area of interest after
making photo-geological studies in conjunction with other available remote sensing
data. If there is any positive indication of mineralization, systematic geochemical
sampling along with ground geophysical survey should be carried out. Once some
anomalous zone are indicated in the area giving some tentative configuration and
extent of the causative body, subsequent step should be done for proving of the area
by pitting, trenching and drilling for collecting samples from the weathered lode
zone and unaltered base metal sulphide if any, to collect more data within host rocks
for analyses. This will indicate strike continuity of mineralized body, if any. Then
detailed mineral exploration is conducted by the methods already discussed.

This type of modelling requires huge amount of surface and subsurface data.
After proper documentation and storage of the data, computer-based classical statistics may be done along with characteristic analyses (after ranking the individual
variables), step-wise regression etc. (Botbol, 1971; Mukherji, 1996) followed by
calculation of probability index for generating predictive contours of the element of
interest for the specific deposit under study. All quantitative data for any deposit
used for the purpose of 'Deposit Modelling' should preferably contain geographic
coordinates of the samples so that all these data may be geographically referenced
as mentioned earlier in order to generate thematic maps whenever required using a
GIS package. Modelling of the deposit at this stage helps in predicting the resource
potential of the study area.

In order to carry out deposit modeling, both qualitative and quantitative variable/
parameter pertaining to the deposit based multi-disciplinary data including assay
values, reserve figures etc. should be considered. Kumari (1996) carried out resource
evaluation of bauxite deposit of Durgamanwadi mines, Kolhapur district, Maharastra.
Initially she carried out classical statistical analyses with the surface and subsurface
data available from the mines including the chemical analysis data of drill cores. In
this area data pertaining to a number of vertical drill holes within bauxite cap rock
above the basaltic rock were available. The results of univariate and bivariate analysis including frequency histogram plots with $Al_2O_3\%$, $Fe_2O_3\%$, and $TiO_2\%$ conform
to a normal distribution. Generally the correlation coefficients observed between
different combinations of variables showed some interesting relationship. The
bauxite derived from basaltic flows are generally ferruginous and its alumina content tends to decrease with increasing iron content and vice versa. The distinct negative correlation between Al_2O_3 and Fe_2O_3 is observed because the colloidal solution,
which deposits bauxite, gets depleted in Fe_2O_3. Another notable feature is the
increase in titanium content showing direct relation to the alumina content in bauxite of basaltic origin. In order to determine the total reserve of bauxite in terms of

grade and tonnage, the drill hole data were used and the block under study was sliced into a number of regularly spaced horizontal segments. Then a global estimate of krigged mean and its 90% geostatistical confidence limits was calculated to arrive at the total resource potential of the area under study.

In this study bore hole data of 91 vertical holes were considered. A series of grade-tonnage relations have also been computed to establish grade-tonnage curves which indicate an ore potential of 3.1 million tonnes of ore at the cut-off of 45.6% Al_2O_3.

Thus, the integrated approach to resource evaluation of Eastern Plateau, Durgamanwadi bauxite mine reveals the use of statistical and geostatistical techniques based on which mine design, planning and economic decision may be outlined. It appears that a sound geological knowledge in combination with statistical and geostatistical modelling techniques provide the foundation of a convincing deposit appraisal. This type of geostatistical modelling for resource appraisal may be used for iron, manganese, chromium, base metals etc. Mukherjee (2003) superposed a network of cells each one square km area on the geological map of Singhbhum copper belt extending from Baharagora in the south-east to Tamadungri in the west using a total of 122 variables. Using statistical analysis as a tool he indicated a possibility of some new mineral deposits in the unexplored areas between the known deposits and in adjacent areas.

2.2.3 Ore Body Modelling

Most of the known metallogenic belts of India have a number of reported deposits with well-defined lode zones, where there is no dearth of different types of surface and sub-surface geoscientific data. Once an ore body with appreciable strike length and width is intersected in different levels (specially in the case of base-metal, gold, platinum etc.), ore body modelling can be initiated provided sufficient subsurface assay data from surface and underground drilling, channel sampling etc. are available (Saha et al., 1986). Then ore body modelling can be taken up. The proper description of an ore body, i.e. its body geometry in 3-D is the foundation upon which follow up mine decisions are undertaken. An ore body has three distinct components, viz.,

i. the physical geometry of the geological units which formed and host the ore body;
ii. the attribute characterization in terms of assays and geo-mechanical properties of all materials to be mined; and
iii. the value parameters in terms of economic mining of the ore body.

The ore body model of a deposit is a representation of its configuration, constructed by interpolating between sample points and extrapolating into the volume beyond the sampling limits. The modelling depends on considerations such as sampling methods, reliability of data, specific purpose of estimation and required accuracy. The basic concept of ore body modelling is to conceive the entire ore body as an array of bricks arranged in a three-dimensional X, Y, Z grid system (X representing easting, Y representing northings, and Z representing difference of elevation from mean sea level) by making suitable assumptions about the continuity of ore body parameters. Each brick of uniform size represents a small block of material to which the values of width, grade, tonnage and other geological entities are assigned (Biswas, 1997).

Prior to ore body modelling, it is necessary to construct a geological model, which would divide the ore body under consideration into as much homogenous zones as possible. With the developments in the recent past, number of modelling techniques are available, which can broadly be grouped into three major types, viz. conventional, classical statistical and geostatistical. Most of these techniques exist as computer packages in some form or other to provide high speed computation of a large number of blocks.

2.2.3.1 Conventional Techniques

These can be grouped under two categories, namely non-mathematical and mathematical. The former can be again divided into polygonal, triangular, sectional, random stratified grid and contouring.

In the non-mathematical models, the value of a sample is extended halfway to the next sample irrespective of the nature of the deposit and mineralisation.

A common practice is to choose a distance, which has been considered in similar type of deposit having sufficient number of samples to overcome the error due to overcome estimation.

Distance weighting methods became more popular when computer assistance becomes available because of the requirement of large number of repetitive calculations. The objective of distance weighting methods is to assign a grade to a block or a point based on a linear combination of the grades of surrounding sample points. Since it generally can be assumed that the potential influence of grade at a point decreases as one moves away from that point, thus grade change becomes a function of distance.

Mathematical trend surface fitting under mathematical model has expanded since its introduction some 50 years ago. The method attempts to fit a mathematical function—a polynomial, to the assay value in a deposit so that the value at any point can be estimated. The confidence of the best-fitted mathematical function is selected by the 'goodness of fit'.

Another mathematical technique is the Moving Average (MA) which produces a trend surface and represents a smooth picture of grade variation but not confined by a mathematical function. It was first used in South Africa to establish block grades that provided the basis for the development of geostatistics which will be discussed later. The method differs from others in that all data surrounding a block is used to value it but once all the blocks have been valued, the point values are deleted for any further calculation.

So, conventional methods involve the combination of sample data with respect to the position of sample, with an intuitive notion of area of influence. These models are unable to define the precision of estimate, which leads to subjective mineral appraisal.

2.2.3.2 Classical Statistical Techniques

The modelling in this technique involves random observations of independent individual of a given population, regardless of the spatial position. This is carried out in order to: (i) produce estimates with specified confidence limits; (ii) provide a check or bias which may have been introduced by conventional methods; (iii) establish average deviation of the observations in any distribution, from its average; and (iv) determine the nature of distribution.

The classical statistical techniques are mainly of two types, normal distribution model and log normal distribution model.

Classical statistical models can define the precision of an estimate, but it has few drawbacks. It is assumed that the samples taken from an unknown population are randomly distributed and are independent of each other. In the case of mineral deposits, this implies that the relative sample positions are ignored and that all samples in the deposit have an equal probability of being selected. The likely presence of trends, zones of enrichment or pay shoots in the mineralisation are all neglected (Rendu, 1981).

2.2.3.3 Geostatistical Techniques

This modelling method utilises an understanding of the spatial relations of sample values within a mineral body. The geostatistical modelling techniques are based on a set of theoretical concepts known as the theory of regionalized variables developed by Matheron in 1971. Any variable which is related to its position and volume (known as support) in space is called regionalized variable. From this theoretical basis, a range of practical methods gradually developed, known by the general term 'Kriging' for estimating point values or block average from a finite set of sample values (David et al., 1987).

Geostatistical methods utilise an understanding of the inter-relations of sample values within a mineral deposit and provide a basis for quantifying geological concept of:

- an inherent characteristic of the deposit;
- a change in the continuity of interdependence of sample values according to the mineralisation trend; and
- a range of influence of the interdependence of sample values.

None of these properties are taken into consideration in the conventional or classical statistical methods. Geostatistic, thus, represents a major advance in ore body modelling, resource assessment and appraisal.

Fig. 2.3 An ideal sketch
of semivariogram showing
sill, range and
nugget points

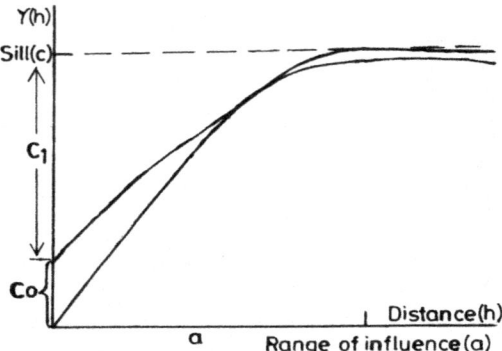

A geostatistical ore body modelling is carried out in two steps, viz. (i) from the geological model, homogenous geological zones are identified which is followed by a geostatistical structural analysis of the sample values by constructing semivariograms (Fig. 2.3) within each homogenous zones; and (ii) a geostatistical estimation by Kriging using the geostatistical structural parameters obtained from the semivariogram after fitting a suitable mathematical function as described below.

The basic assumptions in the geostatistical structural modelling are: (i) values of samples located near or inside a block of ground are most closely related to the values of the block; and (ii) a relation exists among the samples as a function of distance and orientation.

The function that measures the spatial variabilities among the sample values is called 'semivariance'.

The mathematical formulation of a semivariance function $\gamma_{(h)}$ is given by:

$$\gamma_{(h)} = \sum_{i=1}^{n} \left[z(x_i) - z(x_{i+h}) \right]^2 / 2n(h)$$

where $z(x_i)$ is the value of regionalized variable (e.g. grade) at a point x_i in the space, $z(x_{i+h})$ is the grade at another point at a distance 'h' from the point and $n(h)$ is the number of the sample pairs.

A graph, which plots the semivariance values against the lag distances, 'h' is called an experimental semivariogram (Fig. 2.3).

Various mathematical models may be fitted to an experimental semivariogram such as logarithmic model, linear model, spherical model, exponential model, parabolic model etc. In practice the model most commonly used for mineral deposits is the spherical model (Rendu, 1981). Two essential properties of semivariogram are:

1. $\overline{\gamma_{(h)}} = 0$, where $h = 0$.   2. $\overline{\gamma_{(h)}} = \overline{(h)}$ or $\gamma_{(h)} = \gamma_{(-h)}$, i.e. $\gamma_{(h)}$ calculated in one direction should be equal to $\gamma_{(h)}$ calculated in opposite direction. In practice $\gamma_{(h)}$ does not tend to zero when 'h' tends to zero. This is due to the presence of vari-

ance inherent in mineral deposits. This is known as 'nugget variance' or 'random variance', C_o and this phenomenon is known as nugget effect.

The equation becomes

$$\overline{\gamma_{(h)}} = C_o \text{ at } h = 0,$$

Nugget effect may occur due to three principal reasons:

   i. Erratic mineralisation at macro scale, i.e. presence of micro structures;
  ii. Poor analytical precision; and
 iii. Poor sample preparation.

Nugget effect is also referred to as chaotic component that can be considered as the variance of totally random component superimposed on a regionalized variable. The information provided by a semivariogram includes a measure of the continuity of the mineralisation.

The continuity is reflected by the rate of growth of $\gamma_{(h)}$ for small values of 'h'. The growth curve demonstrates the regionalized component of the samples and its smooth steady growth is indicative of the degree of continuity of mineralisation. Sedimentary deposits exhibit high degree of continuity, whereas mineralisations concentrated in veins, veinlets, stringers etc. exhibit low degrees of continuity. At times mineralisation provides no continuity at all. This is known as pure nugget effect model and is common with gold mineralisation.

Point Kriging Cross Validation (PKCV) is the robust method for fitting a mathematical model to an experimental model. PKCV is a technique referred to by Davis and Borgman (1979) as a procedure for checking validity of a semi-variogram model that represents the true underlying experimental semi-variogram and controls the kriging estimation (Sarkar et al., 1995a, b). Based on a crude semivariogram model that is initially fitted by inspection to the experimental semivariogram, estimates of the semivariogram parameters are made from it and cross validated through point kriging empirically.

The geostatistical procedure of estimating values of regionalized variable using the information obtained from semivariogram is kriging. Kriging for point estimate is point kriging and that for a block of ground estimate is block kriging. These are described in details by Isabele Clark (1979) and Rendu (1981). Point kriging is a method of estimation or interpolation of a point by neighbouring sample points using the theory of regionalized variables where the weight co-efficient sums to unity and produces a minimum variance of error. Kriging produces estimates that have minimum error, and also produce an explicit statement of the magnitude of error.

Block kriging is a method of estimation or interpolation of a block of ground with the help of surrounding samples using the theory of regionalised variables. The whole mineralized body is divided into number of regularly spaced horizontal sections, by projecting the sample data from cross-sections. The vertical gap between the sections is kept at length equalizing the vertical lift or bench heights as per the method of mining. In each of the horizontal sections the mineralized boundary delineated on them is divided into smaller grids based on selective mining unit (SMU). Each slice forms a set of X and Y arrays of blocks, with constant Z values (X – eastings, Y – northings, Z – elevation). The array of the blocks are then kriged slice by slice, producing kriged estimate and kriging variance for each of them. The following input parameters are necessary for block kriging: (1) a minimum of 4 and maximum of 15 samples to krige a block and (2) the radii of search for sample points around a block centre to be within 2/3 to full range of influence. The individual slices are then further averaged to produce a global estimate of kriged mean together with associated variance (Biswas, 1997).

Each of the slices are then stacked one below the other from top to bottom, thereby providing a 3D array of blocks distributed regularly in space with their kriged mean and kriging variance and tonnage per block (obtained by multiplying the block volume by the specific gravity of mineral). Such a 3D network of blocks is known as the geostatistical estimation of mineral inventory from which resource position can be deduced.

2.2.3.4 Grade-Tonnage Relations

The mineral inventory is then summarized to establish relations among cut off grade, total tonnage of mineral, total tonnage of waste and waste-to-ore ratio for various probabilistic cut off grades. This is known as grade-tonnage relation and plots of these provide grade tonnage curves (Singer et al., 1975; David, 1980). These relations together with the mineral inventory then form a sound basis for the subsequent mine decisions (Sarkar, 2005).

It is possible to construct all possible types of models for a deposit but the model to be chosen is the one that would be simple, robust and compatible with data. Besides the geostatistical models described here there are also other models, viz. non-linear and non-parametric. While the non-linear techniques include log-normal kriging, disjunctive kriging and multi-gausian kriging, non-parametric techniques include indicator kriging and probability kriging.

The purpose of an ore body modelling is to provide a realistic model based on data whose parameter can be used to estimate block values with minimum variance. The geostatistical techniques provide a basis for quantifying the geological concept of the inherent characteristics of the deposit, a change in the continuity of interdependence of sample values and range of influence of the samples. The technique provides estimation method with smallest possible errors, along with the magnitude of the error. Geostatistics thus represents a major advance in ore body modelling, resource assessment and appraisal.

2.3 MODELLING OF OFFSHORE DEPOSITS

2.3.1 Introduction

So far, Mineral Belt Modelling, Deposit Modelling, Ore Body Modelling etc. have been discussed in details for the terrestrial mineral deposits. Coming to off-shore deposits, these are generally confined to continental margin including the present day sea shore, which marks the region of transition between the continent and deep ocean (Fig. 1.2). Wide variety of mineral resources is present along this region, some of which have been exploited since ancient past. These deposits are mostly present between the coastal plain/beaches up to the continental shelf. Occasionally, some deposits extend even beyond the shelf to the upper parts of continental slope. Such deposits have diverse sources, and the process of their formation is also different. Mineral deposits which occur on the continental margins are mainly divided into unconsolidated and consolidated types, which can again be grouped on the basis of their origin as terrigenous, biogenous and chemogenous deposits. Among these, the biogenous deposits are represented mainly by the offshore calcareous deposits, and the chemogenous deposits by phosphorite, barite, shallow water ferromanganese concretions etc. (Ghose and Mukhopadhyay, 1999). Calcareous deposits are formed in areas of high benthic productivity and phosphorite in areas of upwelling and prolific growth of surface organism.

Potentially economic concentration, in both the cases, takes place in environments of highly reduced terrigenous input. Besides presence of manganese nodules of various size in deep ocean basin, another very important type of recent deposits are represented by hydrothermal mineralisation on and within the modern sea floor along the spreading mid-oceanic ridges, giving rise to polymetallic sulphide deposits with appreciable amounts of Cu, Zn, Pb, Co etc. Iron oxide-silica-manganese oxide deposits are also formed as a result of submarine hydrothermalism. Sites of hydrothermal deposits may be of various types due to different types of plate movements which is beyond the scope of this write up. However, from economic point of view, the deposits are indeed rich in many metals, compared to manganese nodules by about 1000 times the mass per unit area and these may be mined some day in future (Ghosh and Mukhopadyay, 1999).

2.3.2 Deposits of Continental Margin

2.3.2.1 Terrigenous Deposits

Any country which is bordered by sea shore along its margin contain coastal deposits. Normally in such country, rivers largely transport the products of erosion of the land and dump these in the sea. The rocks and its constituent minerals from the land are eroded due to mechanical disintegration as well as chemical decomposition. In such a condition, the minerals undergoing chemical weathering loose their primary identity and as such cannot reach the sea in their original form, whereas chemically resistant minerals are carried towards the sea to be further sorted and graded by waves, tides and currents due to size and density differences. As a consequence, the coarser and heavier grains are retained in the coastal areas and beaches, the finer materials moving further downward into deeper water. Thus placer deposits are formed by the mechanical accumulation of chemically resistant detrital minerals along the coastal belts and beaches giving rise to onshore and off shore deposits. Since these detrital materials are derived from land mass these are called terrigenous deposits. Generally such placer deposits contain economically recoverable concentrations of native metals like gold and platinum, gemstones – chiefly diamond – as well as many oxide minerals (such as, cassiterite, rutile, ilmenite etc.), phosphate – monazite, silicates (namely, zircon, garnet etc.) and many other minerals (vide Table 2.1). Gravel along riverbed and quartz sand abundantly present along coastal areas also constitute very important terrigenous deposits.

Placer deposits on modern beaches occur in the form of closely clustered disseminations, stringers, lenses and layers mostly in sands. Because of the lowering of sea level by about 100 m during Pleistocene, the accumulation of river-borne terrigenous material took place not along the present day coast line but in the off shore areas, depending upon the extent of regression of the sea. Many ancient beaches are now submerged and contain buried placers in the offshore areas of the continental shelves. The principal factors that promote the formation of such placer deposits are presence of suitable source rocks, and processes like weathering, transportation and deposition along a river valley. The provenance of such river should have the desired rock types and conditions conducive for the weathering. Composition of the parent rock type decides the likely mineral assemblage of a particular placer deposit. Diamond placers of Namibia (south west Africa) owe their origin to Kimberlitic rocks which are the primary source of diamond. Rich cassiterite placers of Myanmar-Thailand-Malaysia are the result of transportation of cassiterite from Mesozoic granites. Gold has been reported in the beach sands of the Western USA, particularly along Alaska and coastal sands of Canada. These are derived from crystalline basement rocks of gneissic types. Similarly, monazite, zircon, ilmenite and other placer minerals present along the beaches of Kerala are known to be deriverd from the granite-migmatite-pegmatite complexes and high grade meta-sediments of south India.

For the uninterrupted transport of terrigenous material from the catchment areas of the river basins, appropriate hydrodynamic condition along with favourable physiographic set up is necessary. It is further required that the residual concentrations near the source rock and alluvial concentrations elsewhere along the downstream course of the river should also move smoothly towards sea. Thus, higher energy level of the running water is required to ensure the movement of detrital materials all along the transport path, so that the formation of coastal placer deposits may occur easily (Mallick et al., 1987; Rai et al., 1991). Otherwise, if the energy of the transported system drops somewhere in between the course of the river or stream, the deposition of placer minerals takes place in land, giving rise to inland placer deposits. These are also described as alluvial placers or stream placers. Other locales where placer minerals may accumulate are the bends and curves of rivers in addition to the deposition in depression and irregularities on river bottoms in regions of shallower gradients of flow. Geomorphic barriers and traps formed due to neotectonic movements or other reasons also affect the smooth transport of placer minerals towards sea. It can be surmised that during the Pleistocene low stand of sea level, much of the present continental shelves were exposed and the river flowed further down stream into deeper regions. As a result, off shore 'ancient' placers are found on submerged beaches and at places along the paleo river-channel which are also called as 'fossil placers'. It may be mentioned here that the placer deposits described above are the source of REE and Rare Metal which are dealt in great details in Chapter 6.

2.3.2.2   Biogenous Deposits

Continental shelves and banks at several offshore areas of the world, especially around coral islands, are full of calcareous material (coral and shell) in the form of sand. The ideal situation for the formation of shell deposits seems to be high benthic formation together with low terrigenous influx. Richer deposits with higher calcium carbonate contents are the result of concentration and reworking of shell debris due to wave and current action on the sea floor, thereby reducing the non-calcareous material effectively. Recent to sub-recent fossil shells may also occur in abundance in some deposits due to reworking by hydrodynamic processes. The deposits of shell, coral and calcareous sand were mined on a small scale for extraction of lime since earlier historic time. They now chiefly provide raw materials for manufacturing cement and some chemicals, particularly in those areas where on shore limestone deposits are generally not available. Iceland is an important example which exploits offshore shell containing about 80% $CaCO_3$ for making cement. There are many other areas of offshore shell exploitation, such as Galveston Bay in the Gulf of Mexico, San Francisco Bay and Bahama Banks.

In India, calcareous deposits occur both in the western and eastern coasts. Lime-mud deposit has been mapped at water depths of 180 m to 1200 m containing $CaCO_3$ upto 94% in the western coast off Gujarat, which decreases to about 50% in the upper rise. Large reserves of calcareous sand occur around Lakshadweep, a group of coral atolls, submerged reefs and banks off Mangalore. Sand is of good quality, as shown by low contents of silica, alumina and iron. The following ranges of values in percentages have been obtained for samples collected from lagoons, reefs and sub-merged banks: CaO – 49.25-51.47, MgO – 0.73-2.28, Al_2O_3 – 0.16-1.78, Fe_2O_3 – 0.05-0.25, SiO_2 < 0.1, P_2O_5 – 0.04-0.12 and loss on ignition 42.08-46.31.

A survey of three submerged banks around Lakshadweep by Marine Wing, GSI has indicated a reserve of one billion tonnes of calcareous sand, with an average CaO content of 51% (Rao and Shrivastava, 1996). Shell deposits occur in the backwaters of Vembanad Lake, Kerala, where the thickness of the shell is up to 2 m, at water depths of 1.5 to 2.5 m. The shells have $CaCO_3$ up to 98% and the estimated reserve is 2.5 million tonnes. Shell deposits occur in the Pulikot Lake of Andhra Pradesh. There are relict oolites and biogenous sands present at several places on the middle and outer shelves of Indian coasts (Siddiquie et al., 1984; Sen Gupta et al., 1992).

2.3.2.3 Chemogenous Deposits

Phosphorites are the most common chemogenous submarine deposits. The element phosphorous occurs in a variety of forms on the modern sea floor as submarine deposits consisting largely of phosphate bearing minerals. They are commonly found as nodules and concretions, pallets, slabs, encrustations and as phosphatic mud on the continental shelf and upper slope, in water depths upto 500 m or occasionally more. Phosphates are generally cryptocrystalline aggregates of apatite and have the general chemical formula $Ca_5[(F, Cl, OH)/(PO_4)_3]$. Cryptocrystalline apatite is essentially a carbonate fluroapatite having the composition $Ca_{10}(PO_4,CO_3)_6F_{2-3}$. Complicated substitution by several other elements, contributed by organic and detrital matter, may also occur. Generally, the P_2O_5 content of phosphates varies from less than 10% to about 30%, and the fluorine content may be upto about 3%. Trace quantities of uranium, which may possibly be recovered as by-product of the phosphates, may also be examined for possible extraction.

Geologically recent phosphorites are known from many areas along the western sides of different continents, which are typically associated with areas of upwelling and high biological productivity, for example off south-west Africa, off Peru–Chile, off California and western Australia. In these areas the bottom sediments are diatomaceous mud and ooze, with some calcareous sand and mud, rich inorganic and silicious materials. In other areas, phosphates occur off eastern continental margin, as on Blake Plateau off Florida, off Somalia, east Africa, and on Chatham Rise, south-east of New Zealand. The Agulhas Bank deposits off South Africa need to be particularly mentioned because underwater phosphatic nodules were first recovered from this area about 125 years ago during the Challenger Expedition. The deposit cover a large area at depths of 100 to 500 m of water (Siddique et al., 1984), where bottom sediment is diatomaceous mud, associated calcareous mud, sand and silt.

Geoscientists of Marine Wing, GSI have described four different forms of phosphates from a large area of the continental shelf off Chennai, in the coast of India (Rao and Shrivastava, 1996). These are: (i) dense conglomeratic phosphate, (ii) phosphate nodule (1-3 cm), (iii) calcareous algal nodule (0.5-3 cm) and (iv) phosphatic encrustrations. The phosphates show two narrow bands here at water depths of 150-200 m and 350-400 m, where a maximum of 20% P_2O_5 has been recorded, except the algal nodules which are less common having a lower P_2O_5 content upto 17.5%. In the west coast of India off Gujarat, slabs, irregular lumps and nodules occur at a depth range of 475-510 m on the upper continental slope in a lime-mud environment containing phosphatised lime-stone. The P_2O_5 content of the west coast phosphates are much lower than that occurring off Chennai, only the nodular one have higher P_2O_5 up to 18.5% (Rao and Shrivastava, 1996).

Bedded-type phosphate deposits, which should properly be termed as phospho-rites, are found during the Tertiary (mainly Miocene to Eocene) time. These occur as relicts along continental margins such as those off West Africa and some parts off Peru-Chile. Older bedded land deposits of phosphorites mined extensively on land are widely different in ages from the Precambrian to the Mesozoic.

In this write-up we shall be discussing only the modelling of terrigenous depos-its, mainly placers that are derived materials brought from land sources through rivers and tributaries with Indian examples. However, modelling of any stratified body with lateral extension and with depth persistence, reported from any land area of earth, can be done with the help of computer-based techniques. The beach and the offshore heavy mineral placers along with precious metals like Au, Sn etc. are the major representatives of the terrigenous group of deposits as mentioned earlier. Quartz-rich sand and gravel on the beaches, and in the offshore regions are also included in this group, which are important commodities as construction material, being exploited in many parts of the world. A country like India, having a longer coastline, stands a better chance to possess many potential economic resources in its coastal zone (Fig. 1.3) for exploration. Table 2.3 shows the area and length of coast line of some countries. But it should be remembered that coastal zone represents a fragile ecosystem. So, offshore dredging of the resources should therefore be car-ried out with enough care, so that the coastal physiography and marine biological balance are not seriously disturbed. Exploitation of coastal and shallow sea resources demands regular monitoring of the ecosystem for better management of these resources.

Table 2.3 Area and length of coastline of some countries having potentially important offshore deposits (after Ghosh and Mukhopadhyay, 1999)

Country	Area (sq. km)	Coastline (km)
Indonesia	2027,083	80,791
Australia	7692,300	25,760
USA	9363,396	19,924
India	3287,782	6539
Malaysia	332,556	4675

2.3.3 Indian Beach Placers and Their Offshore Extensions

Most interestingly, features of the Indian beach placers, as such, in its coastal areas are lying to the southern part of the land alignment with the sea. It has a unique geographical location among the countries bordering the Indian Ocean in respect to its overall central position. The sea floor and continental margins, bordering the Indian Ocean covered by a wide variety of terrigenous, biogenous, chemogenous and authigenic mineral deposits of various dimensions, together form the marine mineral resources of this part of the globe. The heavy mineral placer deposits form an important part of this resource. The beach (onshore) and offshore placer deposits of the countries bordering Indian Ocean represent some of the largest non-living marine resources of the world. In this respect, the occurrence of rich heavy mineral placer deposits along certain stretches of both east and west coasts of India is noteworthy (Mahadevan and Sriramdas, 1948; Brown and Dey, 1955; Mallick et al., 1987; Sengupta et al., 1990; Rao and Shrivastava, 1996; Talapatra, 1999).

The northern Indian Ocean region is subjected to the influence of the reversing monsoon every year due to its peculiar global position, and this is true for the coastal region of India too. Annual variations in wind stress, evaporation, precipitation and coastal run-off of this part of the Indian Ocean produce typical ocean current circulation generally not found elsewhere. This has resulted in the formation of significant heavy mineral placer deposits along the coastal areas of India (Table 2.4).

Heavy mineral deposits of India are mostly present as onshore beach placers along the well disposed long continuous beaches scattered throughout its east and west coasts. The offshore extension of these heavy mineral placers deposits have been established by systematic geoscientific surveys and detailed studies undertaken by the research vessels (RV) of the Marine Wing, Geological Survey of India along the coastal areas within the territorial waters. Computer-based processing of the data generated by the different cruises discussed here, may be fruitfully utilized to generate 3D models of the heavy mineral distribution patterns along the coastal areas for placer mineral evaluation and prediction of offshore targets from known to unknown tracts. Percentage of important heavy minerals in placer deposits from the eastern and western shelf differs markedly depending mostly on the varying litho units present in their respective provenance. In the eastern shelf there is a common enrichment of garnet, sillimanite, zircon, apatite, in addition to ilmenite and monazite, while in the western shelf ilmenite, magnetite along with pyroxine, hornblende and muscovite are generally abundant excepting in the coast of Kerala where mixed heavies with garnet, ilmenite, monazite, zircon etc. are abundant (Kumari et al., 2015).

Table 2.4 Physical properties and composition of some placer minerals and metals (in order of increasing specific gravity)*

Mineral	Colour	Sp. gravity	Hardness	Composition	Remarks
Sillimanite	Colourless, yellowish	3.2	6-7	Al_2OSiO_4	
Diamond	Colourless	3.5	10	C	Gemstone, inferior varieties have various industrial uses.
Kyanite	Blue, white	3.6	4-7	Al_2OSiO_4	
Garnet	Various, commonly pink, red, brown or black	3.5-4.3	6.5-7.5	$(Mg,Fe,Mn)_3$ $Al_2Si_3O_{12}$	Moderately resistant to weathering. Used as gemstone and abrasive.
Rutile	Reddish brown, black	4.2	6-6.5	TiO_2	Used in pigment, paper, metallurgy, etc.
Ilmenite	Black	4.5-5.3	5-6	$FeTiO_3$	Weakly magnetic.
Magnetite	Iron-black to brownish black	4.5-5.3	5.5-6.5	Fe_3O_4	Strongly magnetic.
Zircon	Colourless, brown	4.6-4.7	7.5	$ZrSiO_4$	Radioactive when Zr replaced partially by U.
Monazite	Yellowish to reddish brown	4.9-5.3	5-5.5	(Ce,La,Th) PO_4	Radioactive, presence of thorium replacing U makes it a mineral of great strategic value for nuclear power generation.
Cassiterite	Brown, black	6.8-7.1	6-7	SnO_2	An ore of tin.
Platinum	Silver white	14-19.7	4-4.5	Pt	Occurs together with palladium
Gold	Yellow	14.6-19.3	2.5-3	Au	Native gold is usually alloyed with silver (electrum). During transport, Ag being more soluble than Au is leached out. So beach placer gold is purer.

*Modified after Ghose and Mukhopadhyay (1999)

2.3.3.1 Classification and Processes of Development

Placer deposits of India may be defined as a special type of clastic sediments, generally unconsolidated or semi-consolidated type with economic grade concentration of one or more valuable resistant and dense minerals. Formation of such deposits are controlled by the hydro-dynamics of sea waves along the coastal region, as conditioned by the prevailing tectonic, geomorphic and climatic conditions which are valid for all placer deposits (Kunzendorf, 1986). As a result considerable portion of global supply of native gold, platinum, diamond, rutile, ilmenite, zircon, monazite, cassiterite etc. are recovered from placers. Mention may be made here about Chapter 6 where REE and Rare Metal occurrences of the earth have been discussed in details with their techniques of exploration. Classification of placers are mainly based on medium of transport, the environment of deposition and preservation, the distance of transport from the source provenance and the physical property and composition of the heavy mineral. Residual placers are formed immediately above the bed rock on a flat ground by dissolution and/or removal of lighter rock material. Elluvial (or colluvial) placers generally are formed due to downslope creep of disaggregated mass of residual materials by water action. Aeolian placers are usually reworked, second-generation products of earlier beach placer due to wind action. Fluvial placers are generally derived from distance sources due to river or stream action, which may be grouped into flood plain, sand-bar, buried placer etc.

Mechanism of placer formation along the coastal areas of India requires an understanding of the process of transport and deposition of clastic sediments in different geological environment. India being a vast country with various geomorphic features, the combined action of the onrush and backwash, aided by strong beach-parallel current within the surf zone, produces strong movement of sand along the beach of both east and west coasts of India. These interlinked mechanisms are capable of producing rich concentration of placers as observed along the beaches. However, the mechanism of development of placer deposits along the coastal areas is still imperfectly understood, which requires a thorough understanding of various interlinked processes mentioned above.

2.3.3.2 Optimal Condition of Preservation of Placers

Four most common attributes of placer minerals are high specific gravity, high malleability, toughness or hardness and resistance to chemical weathering. The depositional environment for placers must offer opportunity of reworking i.e. these require low subsidence rate of the basin, high depositional energy, protracted crustal stability and moderate sedimentation rate which are all responsible for the development of very rich deposits along certain parts of Indian coast. The sediments that are being carried by river system from the sub-continental catchment areas are brought to the sea water level and heavy minerals are deposited close to the seashore while the lighter minerals carried away from the confluent areas to the deep sea. Shorelines progradation and lowering of sea level are other factors that aid to preservation of beach placers.

2.3.3.3 Sediment Characteristics Derived from Indian Sub-continent

The offshore sedimentation pattern in the Bay of Bengal and the Arabian Sea is mainly controlled by the unique physiographic set up of the country. The Himalayan ranges on the north and the Eastern and Western Ghat mountains running parallel to the east and west coasts of Indian Peninsula respectively, regulate the flow of numerous river systems that carry the sediments load from their catchment areas. Shelf sediments of the marine basin bordering India range in size from sand to clay (Siddique, 1967). The deeper portions are covered with different types of oozes. The various river systems of India carry enormous quantity of detritus, which are deposited partly in the shelf area and partly passes into the deeper sea. In general, sediments up to Middle Miocene age are encountered in the Ganges cone, while the Indus cone consists of sediments up to Late Miocene age (Mallick, 1983).

The mean percentage of heavy minerals from the eastern and western shelves, and deep sea sediments show notable differences in proportions of certain minerals. In the eastern shelf, there is an influx of garnet and sillimanit (derived from khondalites of Eastern Ghats), zircon (derived from charnockites of Eastern Ghats) and apatite (derived from granite gneisses of Orissa and Andhra Pradesh), in addition to ilmenite, monazite etc. (Sengupta et al., 1990, 1992; Rao et al., 1992). Detailed study of these sediments has identified several mineral assemblages characteristic of a particular river regime. For example, the sediments of the Hooghly river basin have been identified as having similarity in mineralogy with the Siwalik assemblage of the Himalayas, which is the major source of sediments in the Bay of Bengal. In the western shelf, there is an abundance of pyroxene, hornblende, muscovite etc., derived from the Deccan Traps, amphibolites, gneisses and pegmatites spread over Maharashtra, Karnataka and Kerala (Mallick et al., 1976).

2.3.3.4 Offshore Placer Deposits of India

Marine wing of Geological Survey of India has conducted detailed study of coastal heavy mineral studies along with offshore placer deposits including REE and RM deposits within the Exclusive Economic Zone (EEZ) of India during the last fifty years with the help of two coastal vessels, which will be described in Chapter 6. Before describing the Indian offshore placer deposits it may be mentioned here that in general heavy mineral placer deposits are observed along the passive continental margins in different parts of the world. This is true for Indian subcontinent also, where both east and west coasts are marked by the presence of number of onshore placer deposits along the beaches. Starting from east, important occurrences of offshore heavy mineral placers in parts of coastal West Bengal, Orissa, Andhra Pradesh and Tamil Nadu have been reported (Mallick, 1968, 1981; Mallick and Sensarma, 2009; Mallick et al., 1987; Sengupta et al., 1990, 1992), while the occurrences along the west coast in parts of Kerala, Karnataka and Maharashtra are also noteworthy (Mallick, 1972, 1986; Mallick et al., 1976; Senthiappan et al., 1987; Nambiar and Unnikrishnan, 1989; Siddiquie and Mallick, 1972). The offshore extension of the onshore heavy mineral bearing placers has been proved (Mallick, 1974) in most of these occurrences by the detailed geoscientific study of samples collected from the shelf zone, followed by sampling during the cruises undertaken by the Marine

Wing, GSI. Some of the most important onshore and offshore heavy mineral placer deposits reported so far along eastern and western coasts of India are shown in Fig. 1.3. Mallick (2002, 2007) suggested various techniques of heavy mineral exploration along Indian coast and listed the resources of Indian coastal and offshore sediments (vide Mallick, 2015) which highlights the presence of REE bearing monazite etc. that will be discussed in detail in Chapter 6.

The inner shelf along the coastal tract within the territorial waters of east coast of India between Digha in West Bengal to Sonapurapeta in Andhra Pradesh has been surveyed on board R.V. Samudra Kaustabh (Fig. 1.4). In the first phase of work, extensive grab sampling along with bathymetry and subbottom profiling was carried out to establish the sediment distribution pattern and surfacial heavy mineral concentration. A persistent sandy layer has been reported between Sonapurapeta and Puri extending from shore upto water depth varying from 30-50 m. This sandy zone in general contains appreciable quantity of heavy minerals comprising ilmenite, sillimanite, garnet, zircon, rutile and monazite (Sengupta et al., 1992) which contain considerable amount of REE. The total heavy mineral concentration in this part of the shelf within the sand zone is reported to vary between 1.54% and 30.85%. Towards northeast from Puri, this sandy layer tapers out and continues further as a very thin strip along the shore. Based on the results of first phase of work, systematic vibrocoring was done in the second phase between water depths of 10 m and 30 m along Sonapurapeta-Gopalpur-Chhatrapur-Digha sector to delineate the sand body geometry and variation of heavy mineral concentration in space (Sengupta et al., 1992).

Studies along the inner shelf off Sonapurapeta to Kalingapatnam also identified rich offshore heavy mineral placers in continuation to beach placers (Mallick, 1968, 1981; Rao et al., 1992). The heavy mineral suite in this part of the shelf (Fig. 1.4) is characterized by ilmenite, garnet and sillimanite with minor amount of magnetite, zircon, monazite, rutile, ortho- and clino-pyroxene, hornblede, epidote, and kyanite. Here, the near-shore heavy mineral-rich zones are generally predominated by ilmenite and sillimenite, while in the offshore beyond 20 m isobar garnet and ilmenite are more abundant. The concentration of ilmenite, garnet and sillimenite in the offshore surface sediments of this area varies from 0.02-16.54%, 0.01-11.96% and 0.01-5.02%, respectively (Rao et al., 1992).

Proceeding further south along the east coast, onshore beach placers are reported from the southern part of Tamilnadu for a considerable stretch south of Pondichery (Roonwal, 1986) which are mostly rich in ilmenite and mixed heavies.

The detailed sea bed survey within the territorial waters along the west coast by R.V. Samudra Saudikama has delineated multimineral placer deposits of Kerala, Karnataka and Maharashtra. Potential areas of heavy mineral sands have been located from near the coastal areas of Tamilnadu-Kerala state boundary in the water depth of 4-40 m; these extend towards north up to Cochin and beyond. Heavy mineral content in bottom sediments samples of the offshore area around Trivandrum in the south-west coast of India range between 0.53 and 14.58% by weight, whereas the adjacent beaches have 8 to over 80% heavy minerals at places (Nambiar and Unnikrishnan, 1989). The heavy mineral assemblage is marked by the predominance of ilmenite, garnet and sillimanite with appreciable amount of zircon, monazite, rutile and kyanite in both beach as well as shelf sediments of this area (Fig. 1.3). Marked differences in relative abundance of heavy mineral species are observed in different pockets of placer deposits, which may be due to the effect of selective sorting of detritus grains governed, locally by grain size and density.

Further north, extensive occurrence of sands in the sea bed within territorial waters off Quilon has been noticed. The sands are fine to medium grained, grayish or light brownish. These are composed of quartz with some amount of heavy minerals. Ilmenite and sillimanite are the dominant heavy minerals with minor amount of rutile, zircon, monazite, leucoxene, garnet and some ferromagnesian minerals (Senthiappan et al., 1987). Vibrocore samples from selected areas show the presence of heavy minerals varying from 3-9% up to a depth of 4.6 m. The content of heavy minerals decreases with depth in majority of the core samples. It has been observed that the concentration of heavy minerals is more where the water depth is <20 m.

Detailed studies of the opaque minerals from the shelf areas off Mangalore were taken up with a view to deciphering the character and distribution patterns of the opaque minerals (Mallick, 1972). Here, magnetite, ilmenite and hematite are abundantly present with some amount of goethite-lepidocrocite, rutile and pyrite. The heavy minerals, derived from the catchment areas of the two rivers flowing into the Arabian Sea near Mangalore, were concentrated at the confluence point due to high density. The percentage of heavies gradually falls away from the shore.

Towards north, presence of heavy mineral bearing sand containing TiO_2 in the range of 5-46.3% has been reported in several small bays of Konkon coast from a preliminary survey conducted by GSI in mid-seventies. Later surveys by National Institute of Oceanography (NIO) along this part of west coast indicated presence of ilmenite sands in water depths of 9-12 m with TiO_2 content of 42-57% in some areas (Siddiquie et al., 1984). Subsequent integrated geological and geophysical offshore surveys around Ratnagiri Bays and collection of systematic vibrocore samples revealed the presence of rich ilmenite and titano-magnetite in very fine and coarse silt fractions. Out of the two main important heavy minerals present in these sediments, ilmenite varies from 1-40% and magnetite from 0.8-8%.

Heavy mineral distribution patterns in the northern part of the Arabian Sea off Bombay to Gulf of Kutch indicated the presence of opaques (essentially magnetite), clinopyroxene, epidote, monazite, zircon, muscovite, biotite and chlorite, with minor amount of tremolite/actinolite, garnet, sillimanite, kyanite, rutile and sphene. The assemblage indicates a mixed igneous and metamorphic source with contribution from Deccan Traps and other rocks present in the Peninsular shield area.

2.3.4 Methodology

Offshore placer deposits including Rare Earth Elements (REE) and Rare Metals
(RM) of any country along with its terrestrial deposits of the world require systematic scientific approaches of its methodology. Application of statistical methods for
evaluation and demarcation of mineral deposits along continental areas or offshore
areas of any country has become a common practice in recent years as already
mentioned earlier for prediction of resources in specific study areas (Botbol, 1971;
Agterberg et al., 1972; Benest and Winter, 1984; Talapatra et al., 1986a). These
methods are suitable for estimating the terrestrial mineral resources of a region, in
areas that represent extensions of mineral deposits/belts in land areas, as well as in
virgin areas (Sarkar et al., 1988; Sarkar and Nair, 2002). In this chapter attempts
has been made to outline broadly the methodology of mineral deposit modelling.
To start with, modelling of offshore placer deposits along the coastal areas have
been attempted here. A systematic approach to resource evaluation of heavy mineral placer deposits along the coastal areas off Gopalpur has been attempted after
geo-referencing the study area, using both classical statistical analysis and geostatistical modelling techniques, taking into consideration the geological controls of
such deposits. Geo-referencing of the entire study area was done by scanning
Survey of India map on 1:2,500,000 scale using Lambert Conical Orthomorfic pro-
jection system (Fig. 2.4). The geostatistical modelling generally reveals an

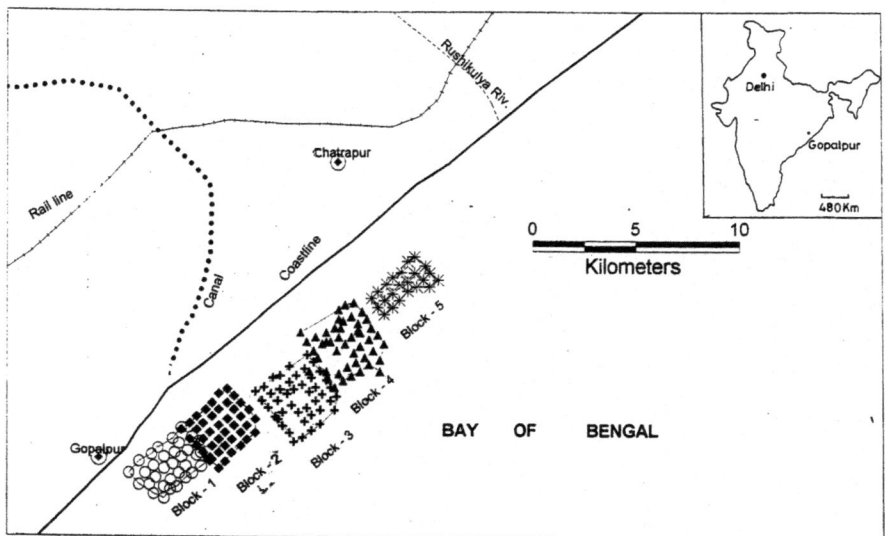

Fig. 2.4 Geo-referenced map of vibrocore sample locations of all the five blocks off
Gopalpur, Orissa

accurate regionalized picture that may be applied to the placer deposits also for heavy mineral placer resource evaluation, where sufficient data at regular intervals are available. This will help to estimate the continuity of placer mineral deposits, both laterally as well as with depth from the available data collected by different cruises.

In order to make out the nature and extent of a placer deposit from the available data, a suitable user-friendly software package may be utilized to produce a colour coded 3-D block diagram of all the blocks of study area showing distribution of total heavy mineral percentage values (Talapatra, 1999; Bandopadhyay et al., 2002).

Most of the common minerals found in beach placers range in specific gravity from 3.5 to 5.3 excepting a few very heavy minerals like cassiterite (6.8-7.1). These are rutile, ilmenite, magnetite, zircon, monazite, garnet, silimanite, barite etc. The active wave action of sea generally extends only up to a few metres of water depth in the coastal areas. That is why the placer deposits generally occur partly along the present day beaches as well as in offshore areas close to the shore, mostly near the confluence of rivers. At times, heavy mineral concentrates may be found as much further offshore or onshore deposits due to past sea level changes, tidal currents, wave action etc. (Kunzendorf, 1986). To and fro wave action of ripples plays an important part in concentrating and accumulating the heavy minerals along the coastal sandy tracts and shelf areas (Ghosh and Mukhopadhyay, 1999). Wind movements may also sometimes produce ripples near the coast resulting into placer deposits, as in the case of tin deposits in the Malaya Peninsula, where cassiterite rich placer was formed by the disintegration of the host rocks of tin deposits.

It has been observed that most of the world's marine placers were formed during the Pleistocene ice age when the sea level was as much as 160 m lower than the present level (Donn et al., 1962). A series of beaches parallel to paleo-coast line and extensive channels were developed during this time along the coastal zones in different parts of the earth, because of the cyclic nature of the sea level changes during glacial and non-glacial periods. This clearly demonstrates evidence of marine transgression and regression in geological past.

2.3.5   Test Application of 3-D Modelling

As already mentioned, Marine Wing, GSI has been collecting offshore heavy mineral data systematically with the help of its coastal vessels along the exclusive economic zone of India. In order to utilise the large amount of data generated by vibrocore and grab samples on offshore area by Marine Wing, G.S.I. (Sengupta et al., 1992) and also to represent this by means of computer-aided graphic modelling depicting the total heavy mineral percentage (HM) distribution pattern, a 3-D modelling technique was attempted to visualize the body geometry of the deposit from the sub-surface data of Block 3 off Gopalpur (vide Fig. 1.5). This shows the locations of vibrocore samples outlining the HM concentration pattern for resource evaluation and prediction of HM placer deposits along adjacent areas (vide Plate 1 in Chapter 1). For this, each vibrocore sample as described in Chapter 1, is utilised. Using this graphic software a 3-D colour coded block diagram of Block 3 was generated (Talapatra, 1999) as per the objective. The 3-D diagram thus generated helped to visualize the body geometry of the heavy mineral placer in the block showing the variation of total heavy mineral weight percentage with depth as well as along its lateral extent (vide Plate 1). The software can generate any cross section both horizontal and vertical for study. It appears from the 3-D block diagram that, in general, top one-metre sediments from 38 cores (out of 46) of the block contain greater than 5% HM while in 28 cores almost similar concentration around 5% have been observed up to a depth of 2 m and beyond 2 m the percentage gradually decreases in this block (Plate 1).

Reference may be made here to the subsequent work done by computer-based modelling of heavy mineral placer deposits off Gopalpur, Orissa, which was taken up by the Marine Wing, GSI with the objective of ascertaining the body geometry of all the five blocks of multi-mineral placer deposit off Gopalpur (vide Fig. 2.4). In this exercise, vibrocore samples of around 4 m length of undisturbed sandy core were collected in a grid pattern roughly (500 m × 500 m) by vibrocorer. In total 191 vibrocore samples were collected from five blocks (Sengupta et al., 1992) and 3-D block diagrams were generated to establish the body geometry of HM placers off Gopalpur (Bandopadhyay et al., 2002). Subsequently all the 191-vibrocore sample data were kept in a data base after geo-referencing with the help of a GIS package (Talapatra and Banerji, 2002) on a research scheme sponsored by DST (Fig. 2.4).

After proper documentation and storage of these data, initially a computer-based statistical analysis like characteristics analysis (after ranking the individual variables), stepwise regression etc. (Botbol, 1971; Mukherji, 1996) followed by calculation of probability index may be done for generating predictive contours of the element of interest (here total HM contents) for the specific deposit under study (cf. Agterberg 1974). Quantitative data for both onshore and offshore deposits stored in the database for processing purpose should preferably contain geographic coordinates so that all these data are geographically referenced in order to generate thematic maps, wherever required, using a GIS package.

The combined 3D-model as suggested in the write up will draw the concentrations of total heavies, extrapolating the intervening gap portions, by the in-built software algorithm. Thus, prediction of heavy mineral resources of the adjacent and extension areas of the blocks under investigation may also be considered after proper validation (Talapatra and Banerjee, 2002).

Another alternative method was also tried with all the vibrocore sample data to show how the HM distribution pattern generate four contour diagrams each for the five blocks of total HM concentration along the seabed level, 1 m, 2 m and 3 m

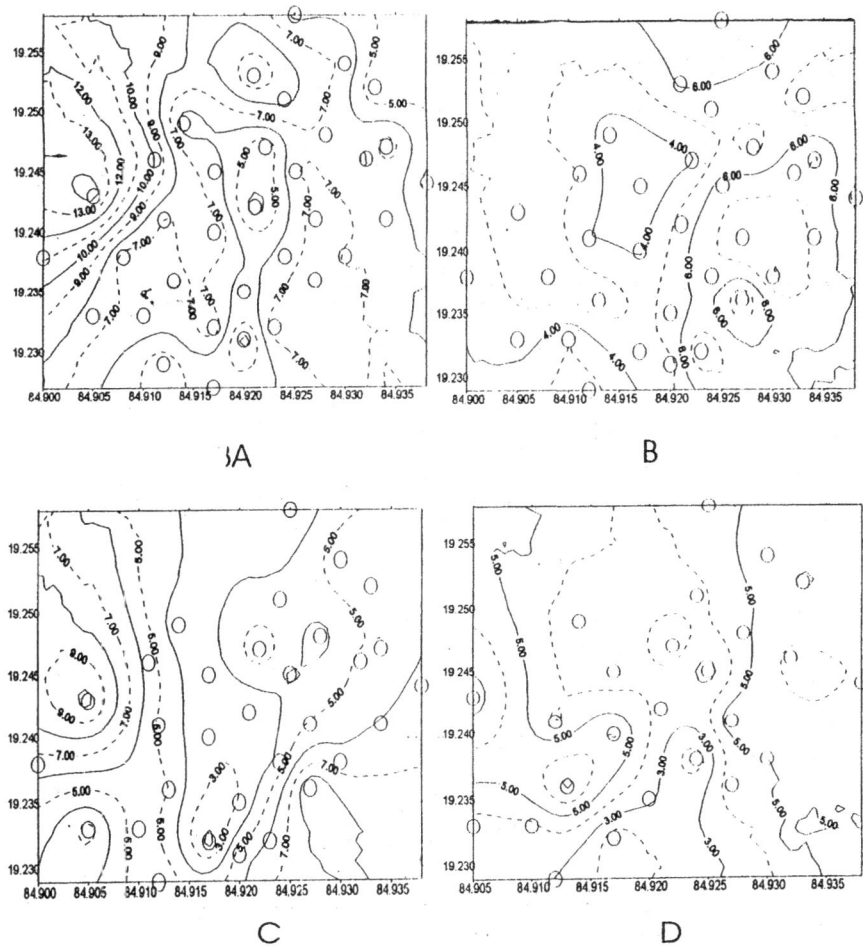

Fig. 2.5 (A-D) Total heavy mineral distribution pattern shown along different levels (sea bed, 1 m depth, 2 m depth, 3 m depth) from Block 1

depth with the help of a graphic software package. It is observed from Figs 2.5A to 2.5D that the computer generated contours of concentration of total HM percentage of Block 1 is generally high in upper levels with some local variation and that higher concentration generally appears to lie parallel to the coast line roughly along NE-SW. Similarly, percentage of total HM of individual Block 2 (Fig. 2.6), 3 (Fig. 2.7), 4 (Fig. 2.8) and 5 (Fig. 2.9) show somewhat similar pattern where percentage is decreasing gradually with depth. But, in Block 5, percentage of total HM

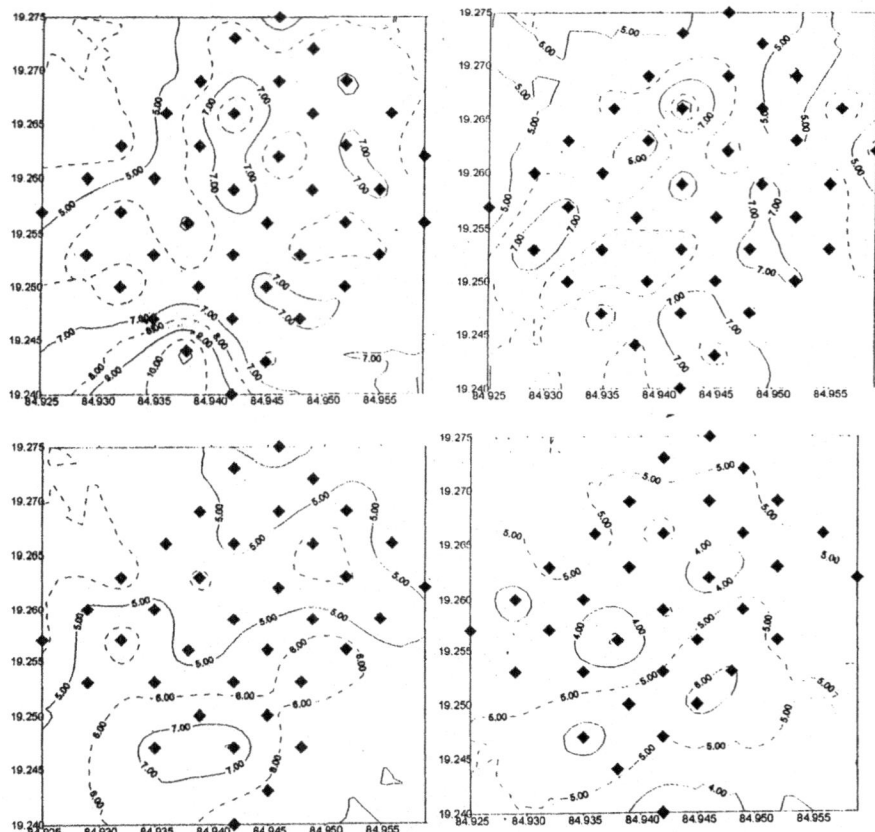

Fig. 2.6 (A-D) Total heavy mineral distribution pattern shown along different levels (sea bed, 1 m depth, 2 m depth, 3 m depth) from Block 2

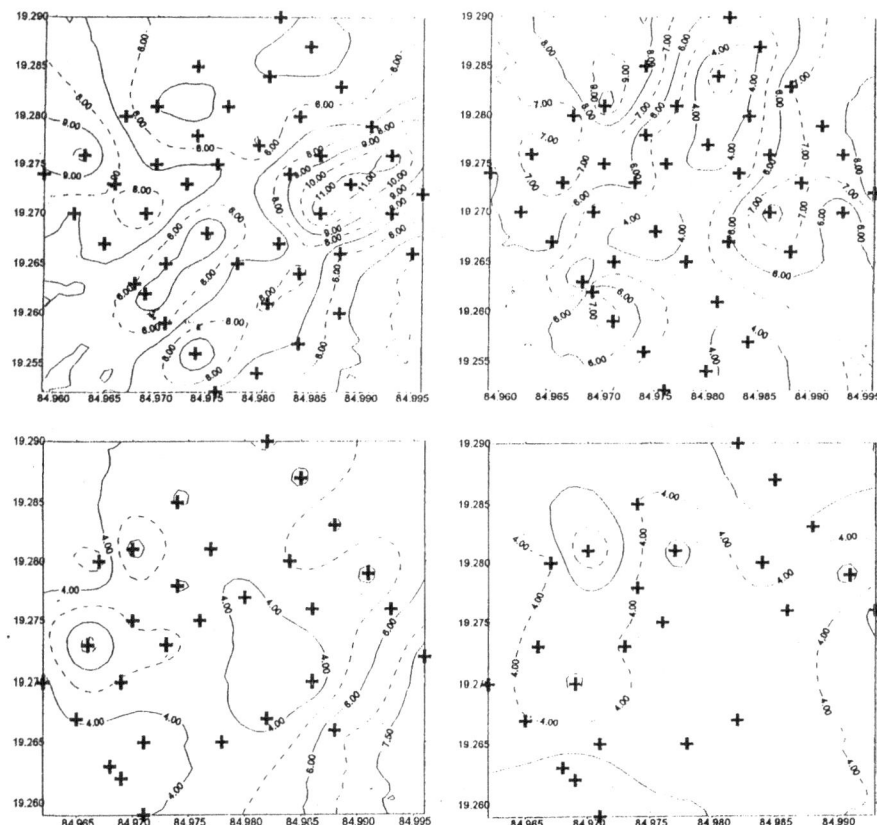

Fig. 2.7 (A-D) Total heavy mineral distribution pattern shown along different levels (sea bed, 1 m depth, 2 m depth, 3 m depth) from Block 3

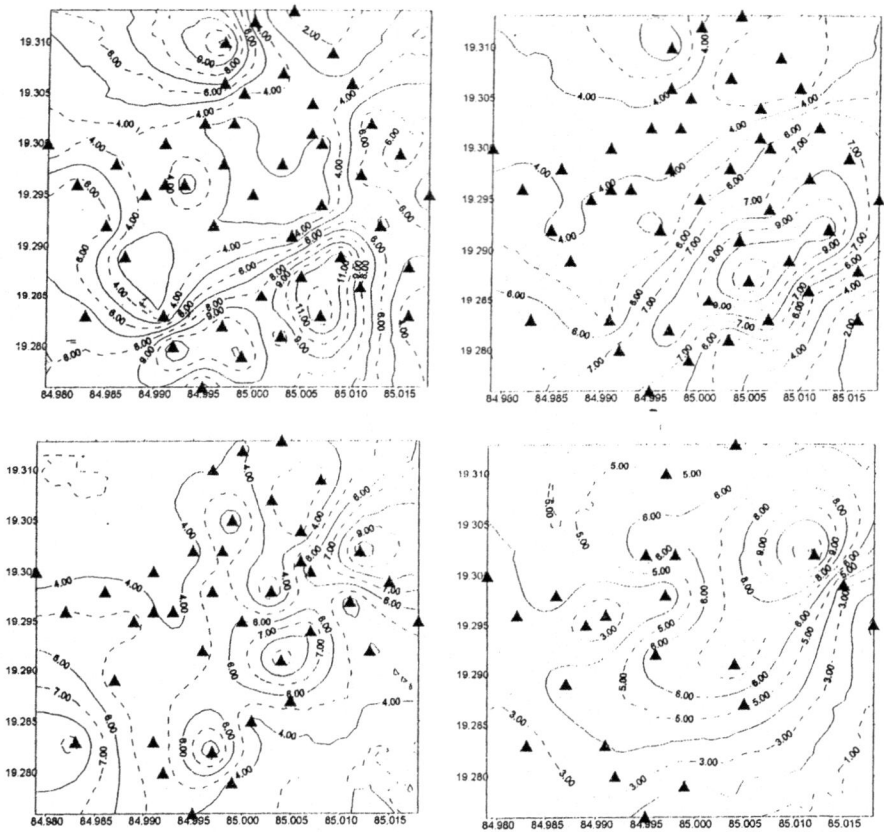

Fig. 2.8 (A-D) Total heavy mineral distribution pattern shown along different levels (sea bed, 1 m depth, 2 m depth, 3 m depth) from Block 4

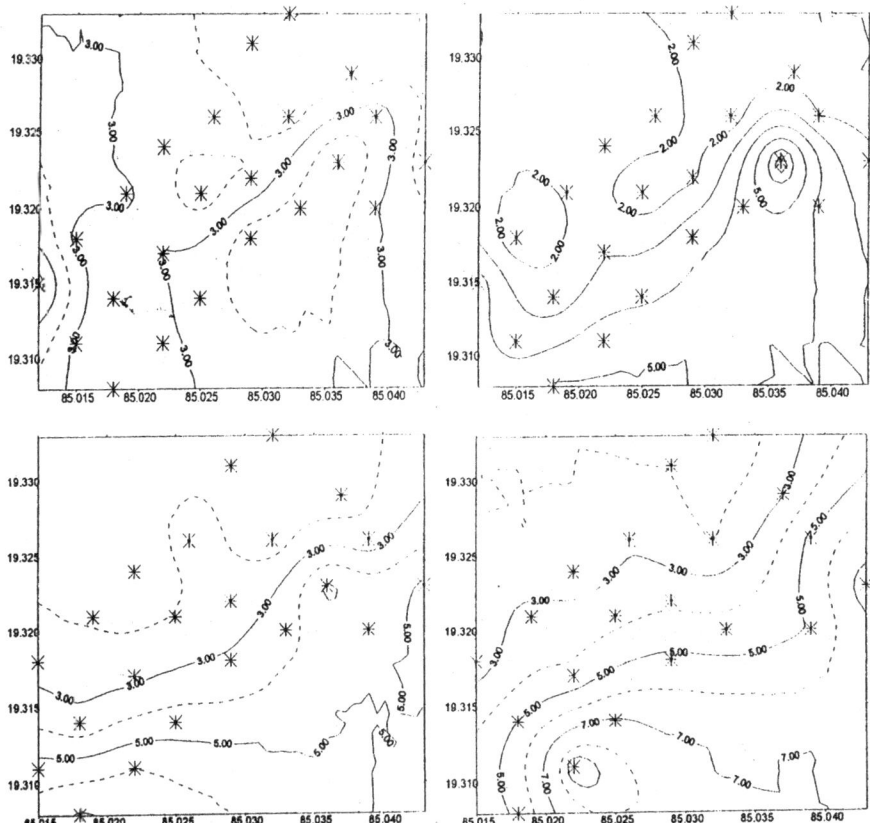

Fig. 2.9 (A-D) Total heavy mineral distribution pattern shown along different levels (sea bed, 1 m depth, 2 m depth, 3 m depth) from Block 5

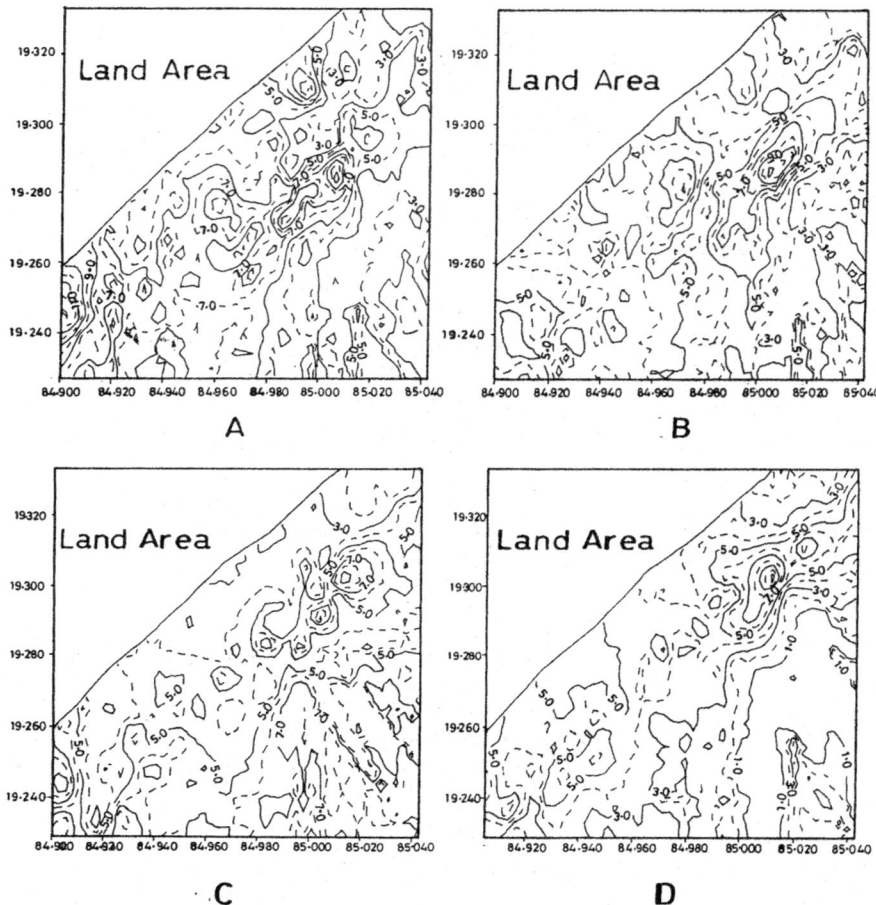

Fig. 2.10 (A-D) Total heavy mineral distribution pattern shown along different levels (sea bed, 1 m depth, 2 m depth, 3 m depth) from the whole area including all the blocks

is very low compared to other blocks, in general (Fig. 2.9). Then the total HM concentration of the entire study area, including all the five blocks (discarding the land portion) were also generated for the respective depth levels, mentioned earlier, which clearly shows the variation of HM concentration of the entire track studied, excluding onshore beach area (Fig. 2.10). This gives a clear conception of 3-D variation of the HM concentration of the study area. In this way, contours can be generated by extrapolating the data of the nearest points, where samples were not collected/available, by the built-in algorithm of the software. Hence actual sampling may be compared with the computer-generated result where there is no sample, to validate in adjacent areas. This computer-based technique will help to extrapolate any known mineral deposit towards unknown areas in any part of the world proving the extension of such deposits.

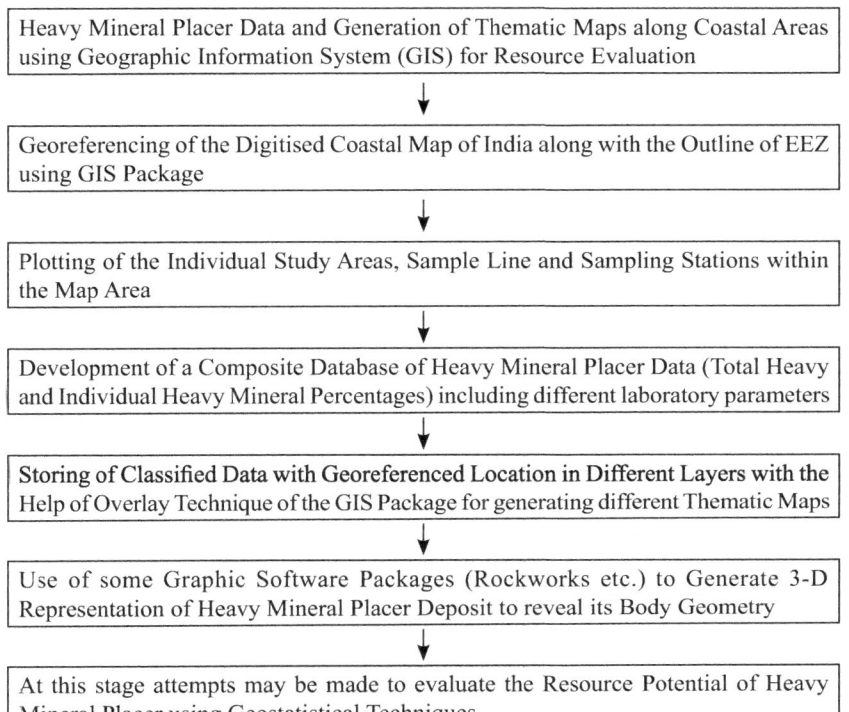

Fig. 2.11 Flow chart showing the procedure of graphic presentation of heavy mineral placer data in 3-D

   From the above paragraphs it is clear now as to how to utilize the vibrocore and grab samples from the offshore areas within territorial waters of EEZ. Once the sample points are geo-referenced and plotted in the coastal map of India up to the EEZ, graphic representation of the heavy mineral concentration within the different depths from the sea bed level is possible. Even a 3-D Block diagram can be generated for the different blocks. This type of study can also be done for any other mineral deposits present in the offshore areas. Figure 2.11 demonstrates the various steps to be taken up in such studies, ultimately for predicting the mineral resource potential of the study area following the different steps mentioned in the diagram.

   It shows how vibrocore samples collected from marine cruises along offshore areas are utilized for graphic representation of heavy mineral placer data. Thematic maps along coastal areas are generated using GIS, plotting the individual study areas to generate the body geometry of the mineral deposit for evaluation of resource potential. As such this technique of utilizing GIS for resource evaluation of any type of mineral deposit in any part of terrestrial or offshore mineral deposits of the world will be beneficial and this may be fruitfully adopted.

Chapter 3
GEOCHEMICAL EXPLORATION OF MINERAL DEPOSITS

3.1 INTRODUCTION

In order to explore the mineral deposits of any country whether exposed or concealed, the basic approach is to prepare the systematic geological map of the concerned area, followed by geochemical and geophysical surveys of the area. In modern times, the geochemical and geophysical surveys have become sophisticated, and the end users multiplied. However, in this write-up only geochemical surveys of Precambrian rocks exposed along the crustal part of Indian continent will be discussed. Geochemistry utilizes the principles of chemistry to explain the mechanisms regulating the variations of rocks, past and present of the major geological systems such as the Earth's mantles, its crust and its atmosphere (Mason, 1966). Just as the geological map (say 1:50,000 scale) essentially depicts the disposition of the rocks of the different geological formations and, depending on the needs, various levels of structural and detailed petrological information can be incorporated in it. Similarly, a geochemical map can aim at projecting surficial chemical variation involving one or more elements. Through a more detailed and advanced study, one might be interested to enquire into the inter-relationship of a suite of rocks in a complex assemblage and their chemical compositional variability. Such an endeavour depends heavily on petrochemical evidences (Rose et al., 1979). Similarly, an environmental geochemist who is interested in studying effect of one chemical element or an association of elements on vegetation or on human life, would like to get the basic data from a geochemical map only.

A geochemical map, like geological, should incorporate the essential chemical data on the various rock units and their secondary derivatives with the ultimate aim of preparing a thematic map of the region. Geochemical maps may be broadly grouped into two types, namely (i) those which show the element concentrations at

© Capital Publishing Company, New Delhi, India 2020
A. K. Talapatra, *Geochemical Exploration and Modelling of Concealed Mineral Deposits*, https://doi.org/10.1007/978-3-030-48756-0_3

the sample locations (point-symbol maps, Talapatra et al., 1984) as shown in Fig. 3.2, and (ii) those which emphasize the regional elemental distribution pattern using contours (Agterberg et al., 1972) in local and/or regional scale. Sometimes additive/subtractive ratios of different elements at the sample points are also used for contouring. Both types of maps normally display a single element for the sake of clarity, excepting the maps generated by various ratios of elements and there are a wide variety of possible display methods of such data (Govett, 1983). Figure 3.1 shows the geological map along three stripes of Chakradharpur-Kharswan area, Singhbhum district where bed rock samples were collected across the western extension of Singhbhum Cu belt (Talapatra et al., 1984). Trace element distribution in ppm has been shown by means of point symbols for Cu, Ni, and Co of each sample points separately as shown in Fig. 3.2 (a), (b) and (c) to establish some geochemical signatures, if any, for defining the western extension of the Singhbhum Shear Zone, also known as Singhbhum Cu belt and its mineral potentiality.

It is interesting to note that moderately high values of some trace-elements were noted from the northern and southern fringe of the stripes of the area studied, which roughly corresponds with the northern and southern extensions of the Singhbhum Cu belt, respectively (inset map of Fig. 3.1). Detailed subsurface data should be collected along these areas to locate the anomaly, if any. In all the cases, good map design should be concerned with contrasting the information of primary interest, the

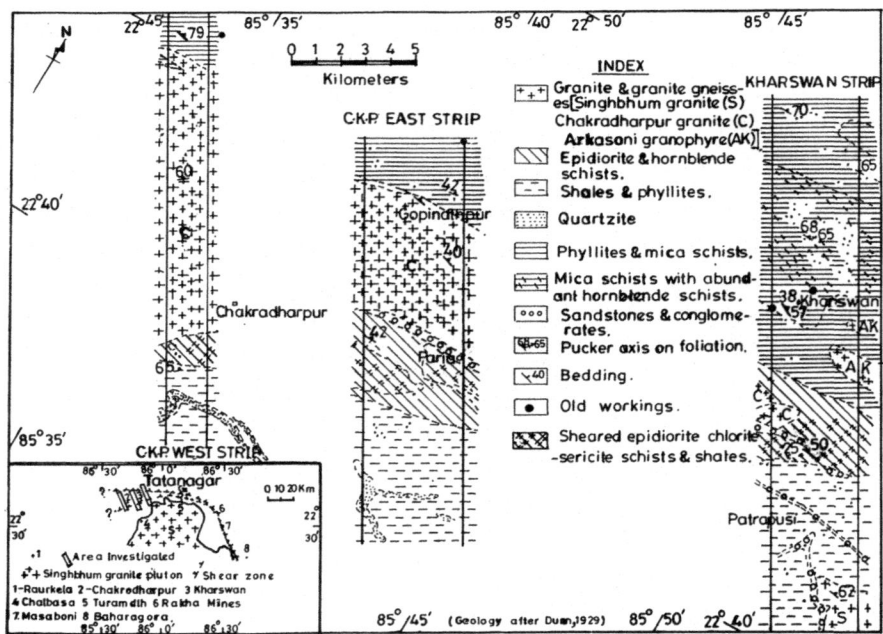

Fig. 3.1 Geological map along the three strips of Chakradharpur-Kharsawan area, Singhbhum district. Inset map shows the location of the Singhbhum Shear Zone

Fig. 3.2 Distribution of
(**a**) Cu, (**b**) Ni and (**c**) Co
within the bedrock samples
of Chakradharpur-
Kharswan and their
concentration represented
by the size of various
symbols on each sampling
point for the different
elements and their
concentrations represented
by the size of symbol

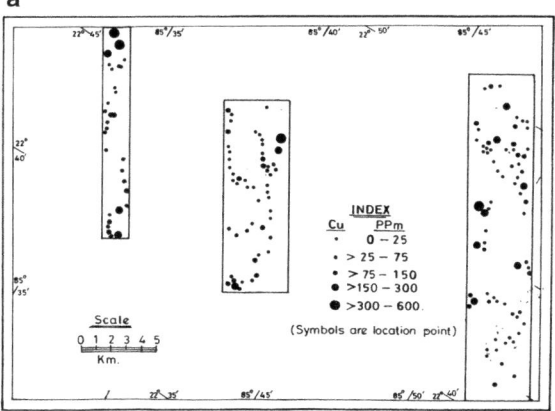

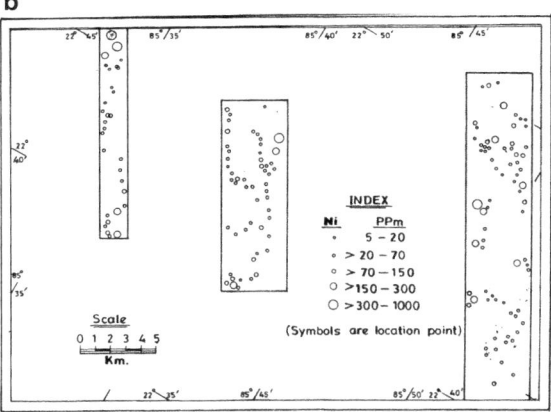

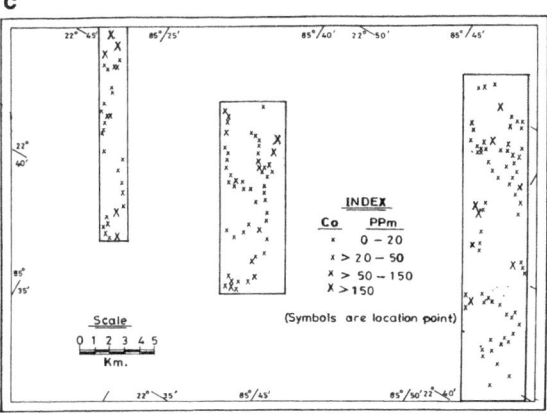

geochemical variation, with necessary parameters on geographical, topographical and geological details. It is often convenient to superpose the geochemical and geological maps for this reason. This has, however, become easy with the development of computer-based overlay techniques using graphics and GIS packages. The details of these techniques are beyond the scope of the present write-up.

The Genetic Aspects of Mineral Deposits

Locating any mineral deposit in any country requires detection of some geochemical anomaly that will point out the presence of some economic mineral occurrences in the area, which may lead to detection of some possible mineral deposits. The most conspicuous event about the origin of ore deposits in the earth's lithosphere is non-uniformity in their distribution showing distinct anomalous zones. This characteristic of mineral deposit specially the hidden ore makes it challenging and requires state-of-the-art techniques to explore them (Ravi Shanker, 2001). At the same time by virtue of this special feature of mineral deposits, it should be possible to locate them in almost all possible geological environment, once the processes of mineral formation are properly understood. This type of conceptual/genetic modelling is valid for both land and off-shore deposits.

Extensive work has been done on the metallogenetic studies of land deposits. This is valid for Singhbhum Shear Zone, India which records copper sulphide deposits along this long shear zone with atomic minerals, apatite-magnetite etc. along this belt. The existing views in regard to metallogenic provinces and deposits are that they can be monometallic or polymetallic, isochronous or polychronous, and isogenetic or polygenetic. They are the products of many compositional and genetic alternatives in the framework of dynamic earth system (Condie, 1982).

It is important that metaliferous deposits and the systems that generate them are viewed not as geochemical chance occurrences. They are, in fact, logical and fortunate culminations of normal geological processes of the earth throughout its long history from very ancient times, which are still continuing in some parts of earth. Quaternary and Recent mineral deposits have been known from very ancient times and during the last decades, deposits have been discovered along the mid-oceanic ridges which have been formed during the ongoing hydrothermal activities related to plate movements of the earth.

Ore deposits can be divided essentially into two major processes, viz. (1) endogenic and (2) exogenic. The endogenic processes are invariably associated with thermal processes, related to tectonic and magmatic events, triggered and controlled by plate movements. Exogenic deposits, on the other hand, are formed by surficial processes e.g. weathering or shallow marine sedimentation. The latter one has indirect relationship with the tectonic environment.

Plate tectonics, which control the movements of plates, represent fundamentally a mechanism by which excess thermal energy of the earth from the mantle flows outwards through conduction and/or convection (Fig. 3.3) effecting the different continental parts of the earth. Tectonic activities involved in this may give rise to large-scale horizontal movements, rotation of plates, hotspot activities etc. It may also involve basin and ridge type extensional tectonics besides subduction and

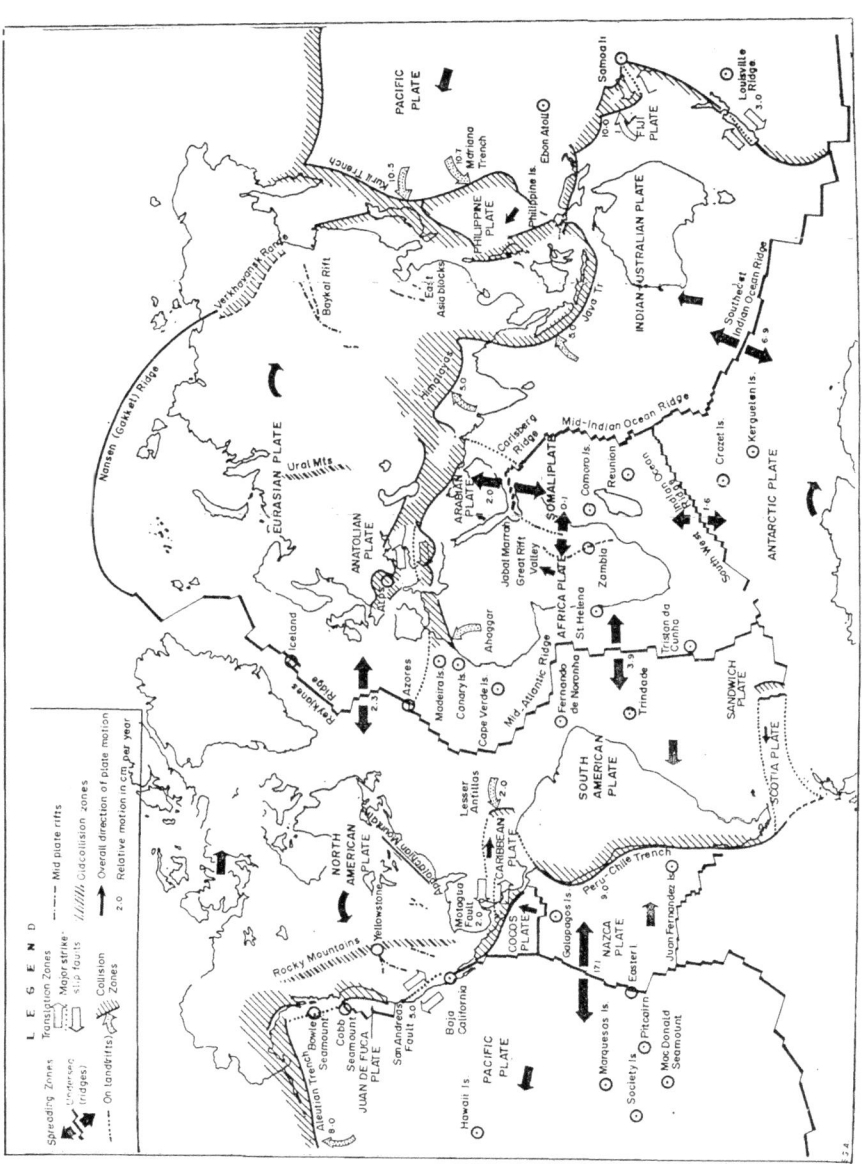

Fig. 3.3 Different plates and their movements are shown in this figure including submarine mountain range, the Mid-Ocean Ridge (MOR), formation of new oceanic crust etc. The circles with dot inside represent important hotspots (After Ghosh and Mukhopadhyay, 1999)

related compressional tectonics at convergent plate boundaries. During the operations of plate tectonics, sometimes it may give rise to metal deposits during Phanerozoic as well Proterogenic times. Burke et al. (1976) put forward arguments that the Archaean-granite-greenstone terrains and certain younger cratons are related to ancient subduction processes.

The bulk of the material that form volcano-plutonic arcs in the plate settings are fluxed from the asthenosphere wedge that overlies the subducting plate. In addition, styles of subduction can vary in terms of rate, angle and thermal condition of the constituents i.e. the plate vectors and imposed stress fields. These variables affect the tectonic, magmatic and sedimentation patterns of extensional, neutral or compressional arcs at all stages of their development. Extensional arcs, for example, tend to be dominated by basaltic and andesitic volcanics and their plutonic equivalents and the related sedimentary fans reflect the mafic source, as in the Andean arc system of Peru and Chile. Compressional arcs, on the contrary, develop thick crust of andesitic-dacitic-rhyolitic volcanic and associated granodioritic plutonic igneous rocks with the production of thick and more felsic sedimentary fans.

Divergent plate boundary environments of the present earth are characterized by the ocean floor spreading often giving rise to mid-oceanic ridges. In the fabric of the continents also, there are terrains related to part rifting, some others having undergone extensive later orogeny, at least in part, due to complete revolutions of Wilson cycle i.e. opening and closing of ocean. In the spreading ridge system, the slow-spreading ones tend to exhibit well-defined axial rifts and greater structural and petrochemical diversity, as compared to fast spreading ridges. These have definite significance in nature and style of metallogeny.

There is enough justification to believe that **mantle plumes** and plate tectonics that might have cause-and-effect relationship, have been operative in the earth system, at least since late Archaeans. A vast amount of melts are generated in the mantle in settings other than convergent plate margins. The generation of melts has important implications for both magmatic and hydrothermal ore deposits (Piranjno, 2000). A direct link of the mantle plumes to mineralization can be found in generation of mafic-ultramafic magmas as in the case of Ni-Cu sulphide deposits associated with flood basalt. There can also be an indirect link of the same with the high geothermal gradients that are established in the crust above the plume, where mesothermal lodes can develop.

Hot-spot activity represents an interaction of lithosphere and underlying asthenosphere in the mantle plume setting and is considered a valid plate tectonic process. Hot-spot volcanic centres are commonly associated with crustal doming with diameters even upto 2000 km and elevations upto 2000 m above the mean sea level. Hot-spot volcanism is not related to plate boundaries as it occurs within the plates. Fragmentation or rifting of continental crust, with formation of triple-rift junctions may be linked to hot spots. Fragmentation of super-continents, like Gondwana land and Rodinia, may have been formed due to mantle dynamism and generation of plumes.

The basaltic volcanism involving flood basalts may occur, as in the case of Deccan plato basalts of western India where hot-spots strike onto overlying continental crust. Crustal melting in response to hot-spot activity can produce anorogenic alkali granites prior to rifting. The volcanic and sedimentary rocks pre- and post-dating rifting fill the rift basins to provide suitable environmental elements for metallogenesis. Rocks in such basins and their metamorphic equivalents are widely distributed in the continents, which host a number of diverse types of metal-iferous deposits related to continental layered intrusions and hydrothermally affected sedimentary rocks.

Convergent plate boundary environment in island arc and continental margin settings are narrow and well defined, characterized by the formation of Cu, Fe, Mo, Au and Ag deposits that exhibit a close association with calc-alkaline magmatism. The deposits are of porphyry type, related to subduction processes, as in North and South Americas, but where the arc system has undergone considerable back arc spreading, as in Japan, the rocks appear to be devoid of porphyry copper. Kesler (1973) has recognized that porphyry copper deposits can be divided into (i) molyb-denum and (ii) gold bearing types. Gold enriched example of porphyry copper deposits tend to be more prominent in island arc settings. Some times a broader zone of chloritic-epidote-carbonate alteration assemblage surrounds the rich zone of metals. Mineralisation in this outer and broader zone may contain Pb, Zn and precious metals in addition to copper.

Stable isotope study by Sheppard and Tailor (1974) described evidences for meteoric origin of the fluids that were involved in vein forming processes in this setting. Roedder (1971) indicated that the metallisation took place at the temperature range of 200 to 300 degree centigrade. Besides development of porphyry copper, copper bearing breccia pipes also develop in this kind of settings as found in Chile and Mexico with low tonnage and high grade copper ores associated with Mo.

Metal deposits of ore-related rifts around convergent plate boundary arcs are quite a common feature. Such regions develop more readily in or at the back of island arc systems in the ocean, as also close to, but not within the continental margin. Several important types of metal deposits generate within such regimes that include 'Climax' type porphyry molybdenum deposits, base and precious metal vein deposits of the Colorado mineral belt and even non-metallic deposits, like that of fluorite in north-eastern Mexico. Major Mo deposits are associated with high-silica, alkali-rich rhyolite porphyries, emplaced as composite diapiric stocks. Kuroko deposits of Japan exhibit a close relationship to submarine dacite and rhyo-lite volcanism. A plate tectonic control on this volcanism has been well established. In Hokuroko district of Japan, there is strong evidence of a period of rapid subsid-ence probably due to caldera formation and accompanying dacite volcanism. The important massive sulphide deposits within this setting developed during the wan-ing phase of this pulse of volcanism.

Divergent plate boundary environment and related metallogeny are mainly formed at mid-oceanic ridge system. Essentially similar oceanic crust is also gener-ated during early stages of continental separation, as in Red Sea environment. Similar feature is also observed during formation of marginal basins in some sub-duction related back-arch regions.

Emplacement of major ophiolites, represented by oceanic crusts in younger oro-genic belts, inevitably involves an arch-continent or continent-continent collision event under normal conditions of subduction. Only thin slivers of ophiolytic materi-als tend to be incorporated in imbricate fore-arch mélange zones. Hydrothermal activity begins at the new oceanic crust, leading to localized green schist to amphib-olite facies metamorphism within it. Low temperature alterations under the influ-ence of convecting sea water that serves as the vehicle for heat away from the ridge zone are characteristics of such geological setup. Where highly permeable regime of oceanic crust overlies magma chamber, metal-rich fluids may reach the sea floor in amounts necessary to form economic ore deposits.

Cyprus-type massive sulphide deposits, as the one in Troodos Ophiolite, can form at submarine spreading centres. The deeper plutonic portions of the Ophiolite complexes may contain economic deposits of chromite, but on theoretical consider-ations such deposits must get transported into the subduction zone and lost if not obducted as an ophilolite mélange. Cyprus-type massive sulphide and podiform chromite deposits are the only hypogene metal concentration of significance in ophiolites. Gold in economic concentration also occurs occasionally in serpen-tinised ophiolitic rocks, as in Ovado, western Liguria, Italy.

Continental rifting and related metallogeny are now locked into the fabric of the continent, which are demonstrably initiated by continental rifting events. These belts represent the settings for rift-related metal deposits like epigenatic hydrother-mal copper deposits as in Zambian copper belt. A rift-related molybdenum deposit as in Oslo Rift of Norway is an example of stratiform copper deposits. Similarly stratiform copper deposits as at Mt. Isa, Queensland, Australia and sediment hosted copper deposit of Zambian type are examples of such mineralisation.

Metallogenesis related to continental rifting has distinct characteristics in response to the different stages of rifting. In pre-saging rifting stage (Mookherjee, 1999), mafic magma generated presumably because of hot-spot activity invades the continental crust forming huge layered complexes locally like Bushveld deposit, South Africa that hosts Cr, PGE and Cu-Ni deposits. Anorogenic granites develop in such situation due to crustal melting giving rise to tin deposits as developed in Jos Plateau of Nigeria.

Metallisation in relation to collision environment is exemplified by Alpine and Himalayan orogenic belts that provide evidences and imprints of Cenozoic collision tectonics. The metallic deposits of collisional orogenic belts can be of two types, one formed prior to main pulse of tectonism and other generated as a consequence of the tectonic and associated metamorphic activity. There is a natural tendency of collision orogens, produced by coming closer of an ocean tract to contain a variety of juxtaposed ore deposit types formed initially in a variety of tectonic settings. Metalogenesis in collision tectonic environment takes place into distinct models. One is significant thickening of the crust especially where Benioff Zone is shallow dipping (as in Bolivia) and generation of crustal-melt granites that eventually gives rise to light ion lithophile, namely, Sn-W and uranium deposits. The other is the collisional events at the continental margins and resultant compression along with dewatering of sedimentary packages to give rise to suitable hydrothermal fluids and other mechanisms for long distance travel giving rise to development of distal deposits of Mississippi Valley type, USA. Granite emplacement in the continental crust of subducting plate, following and perhaps in course of continental collision, is exemplified in Himalayan younger granites. Two-mica-tourmaline bearing Badrinath granite, for example, has apparently been emplaced as a consequence of rise of deep-seated magma from near-vertical Benioff zone. This happened following increase in the angle of descent of the oceanic crust of Indian plate as a result of continental collision. Theoretically this granite, therefore, can be a favourable host for Sn and W.

Geological features controlling a metallogeny in plate tectonic model along lineaments representing fundamental flaws in continental crust are the major factors controlling metallogenesis. Reactivation of the zones of weakness within continental crust gives rise to prominent lineaments, which can be related to the process of development of transform faults in nascent ocean basins within the continents. These sites, therefore, control emplacements of intra-continental alkaline magmatism, as in the western part of Africa and elsewhere. Such settings may not be very congenial for metallogeny, but diamond and, in rare occasions, minor precious metals and rare earth deposit can be associated with this. Many lineaments in Indian subcontinent have close relationship with metallisation, but they are mostly products of orogenesis (Talapatra et al., 1995; Talapatra, 2004). Deccan flood basalts might have some kind of relationship with the Son-Narmada system of lineaments but metallisation related to this system is still a subject of research.

Role of fluids in old formation in the dynamic earth system are plenty as evidenced in deep levels of the crust. The presence of veins in metamorphic rock is a direct evidence of fluid flows along structural and tectonic fabrics. Igneous fluids as modified by the action of metamorphic and meteoric fluids, play important role in formation of hydrothermal ore deposits. While sedimentary sequences contain large quantities of aqueous fluids mainly within pore spaces and fractures, quite a good amount of released fluids originate from increasing pressure and temperature under conditions of metamorphism. It is widely accepted that under-plating of the crust by mantle melts causes a large thermal anomaly. This thermal energy may be responsible for assimilation of metal-rich components from the crust, including greenstone rocks. Induction of the process of degassing of carbon dioxide, water and sulphur from the crust starts at this stage. The volatiles, thus generated, can take up metals in highly mobile complex ions and transport upward and outward for emplacement at the favourable structural and/or lithological sites. Numerous fluid inclusion studies on mesothermal gold deposits corroborate this view.

Ring complexes, many of which are associated with carbonatites, are supposed to be related to intra-continental rift systems and mantle-plume related hot-spots activities. The ring complexes and breccia pipes are products of localized episodic melting, associated with development of large volumes of volatiles, resulting in forceful injection and emplacement towards the surface. These pipes are recognized with their characteristic mineral assemblages.

In Indian context, mineral deposits that we know today, occur mostly in the near-surface situations. Time is now ripe to make a beginning of the large scale concept oriented search using computer-based mineral deposit modelling described in Chapter 2 for concealed deposits using the different techniques of exploration geochemistry which will be discussed in details in the subsequent pages. One can draw a conceptual/genetic model taking into account the occurrence and origin of the deposits. However, study of the geochemistry of the landscape of the area is very important for better geochemical understanding of such deposits.

3.2   LANDSCAPE GEOCHEMISTRY

Importance of geochemical studies for detecting and locating new economic mineral deposits has already been emphasized. For this type of studies, systematic sampling and analysis of important elements from different types of samples like soil, bedrock, stream sediment, plant etc. are essential. Geochemistry is primarily concerned with (1) the determination of the relative and absolute abundance of the elements of the earth, (2) the study of the distribution and migration of the individual elements in various parts of the earth with the objective of discovering the principles governing this distribution and migration, and (3) the application of geochemical principles and information in solving human requirements. Exploration geochemistry, in particular, utilizes the concepts of geochemistry in exploration dealing with the live and interactive processes of elemental migration in the practical field. Such processes largely depend on the original elemental landscape of the terrain (Lovering and McCarthy, 1978) and their paragenetic assemblages as they imperceptibly get modified with elemental movements and fixation.

Geochemical studies are done both on the bedrock, as well as on the stream sediments and soils above them and even on the water, plants etc. contained by the sediments/soils and air above. The elemental distribution which accompanied the ore or rock formation processes fall in the domain of primary dispersion, while the distribution resulting from all other subsequent processes are put under secondary dispersion. It is quite well known that physiography, climate, rainfall and biological processes play important roles in the element dispersion in the surficial domain. Therefore, the geochemical characteristics of the surfaces, especially where bedrock is not exposed, are dependent largely on exogenous processes. Since the major objective of geochemical studies pertains to concealed or blind deposits, bedrock geochemistry may fail due to paucity of exposures and the sediments, soils, water etc. are main media for sampling. In view of these, the concept of landscape geochemistry is evolved for typifying the exploration geochemical conditions in the secondary environment. Physiographic boundaries have a distinct relevance in these studies specially slope of the terrain i.e. topographic features vis-à-vis the dip direction of lithounits and the strategy of study and interpretations require an understanding of the same.

Landscape geochemistry involves consideration of entire volume of exposed rock including the earth daylight surface and extending downward to fresh bedrock. No single medium is considered in isolation. The landscape geochemistry approach was first developed by Polynov (1937) and elaborated in the U.S.S.R. In the seventies there were successful applications in Canadian Cordillera and the Shield. The basic idea is that during geological time under a given set of climatic constraints, the chemical interactions between the lithosphere, hydrosphere, atmosphere and biosphere are largely controlled by relationships between the daylight surface and water table. Accordingly, four elementary landscape types are identified by means of idealized conceptual models of 'landscape prisms': (a) illuvial landscape where the water table is below the daylight surface and evaporation exceeds precipitation, (b) eluvial landscape where the water table is below the daylight surface, but precipitation exceeds evaporation, (c) a super-aqual landscape in a marsh (or bog) where the daylight surface and water table coincide and (d) the aqual landscape in areas of river or lake where a layer of water permanently lies above the solid surface. Application of this concept permits working out of basic principles for the circulation of the geochemical elements in landscape, which may be directly applied to synthesis or analysis of the vast amount of data that has accumulated over the years. The landscape geochemical models indicate the mechanisms of formation of geochemical anomalies only, and not the magnitude and size, the latter being controlled by many local conditions. This mechanism is important in the secondary environment for identifying the efficacy or otherwise of the exploration geochemical methods in a particular area. In all regional geochemical campaigns landscape geochemical approach is desirable.

3.2.1 *Chemical Equilibrium in Surficial Environment*

An understanding of the chemical equilibria that determine the behaviour of the chemical elements in surficial environment is essential for geochemical prospecting/exploration. The mobility of the chemical elements in the surficial dispersion cycle is largely controlled by their solubility in water. Solubility of a given element in water is the relationship between the concentration of hydrogen ions in the solution expressed as pH, and the oxidation potential, or Eh. The former parameter is customarily expressed as the negative logarithm of the hydrogen-ion activity, for which the symbol pH is used (Rose et al., 1979). Natural waters can be grouped into four classes according to the pH value: (a) strongly acidic with pH 3, (b) acidic and weakly acidic with pH ranging from 3.5 to 6.5, (c) neutral and weakly alkaline with pH ranging from 6.5 to 8.5 and (d) strongly alkaline with pH > 8.5. Under natural conditions of a supergene zone, weak acids (CO_2, organic acids etc.) and strong bases (Na, K, Ca, Mg) are prevalent. That is the reason why landscapes exhibiting weak acidity (forest and tundra plain), neutrality (mountainous forest and mountainous tundra), and neutrality and alkalinity combined (steppe and desert) are particularly unevenly distributed. Aqueous solutions with marked acidity can be found only in individual localities, e.g. in and around oxidation zone of sulphide deposits, or in areas of volcanic activity. Similarly strongly alkaline soils generally manifest sodium salinisation.

The other important parameter that controls the mobility of elements is the concentration of electrons in the environment, called the oxidation-reduction or *redox* potential (Eh). This factor is important because many elements occur in more than one valency or oxidation state, and the properties of elements both in solutions and solids change considerably from one valency to other. The concentration of electron is most conveniently measured and expressed as the Eh, a voltage measured between a platinum electrode and a standard hydrogen electrode immersed in the solution (Garrels and Christ, 1965). Generally a high Eh indicates an oxidizing system and low Eh a reducing system. Redox conditions at the ground surface are, thus, due to the dominance of oxygen environment owing to free oxygen in the atmosphere, its dissolution in the waters of the particular landscape and its liberation during photosynthesis of plants. The boundary of an "oxygen level" below which the medium does not contain free oxygen any longer agrees with the ground water table. Reduction conditions prevail when there is an abundance of decaying organic remnants, mainly vegetation, and products of the activity of microorganism.

An indication of the change in redox conditions is the colouration of rocks and overlying soils. Normally red and brownish shades are due to the presence of Fe^{+3} – characteristic of oxidation condition; grayish green and dove-coloured shades – characteristics of Fe^{+2} compounds corresponding to reduction conditions. With alkalinity, and at high temperatures oxidation reactions become easier than in an acidic medium and at lower temperatures. This is specifically what is responsible for the prevalence of oxidation conditions in deserts and reduction conditions in tundra landscapes.

3.2.2   Dispersion

Materials of the earth normally do not maintain their identity as they pass through the major transformations of a geochemical cycle, but rather tend to be redistributed, fractionated or mixed with other masses of material. This process, in which atoms and particles move to new locations and geochemical environment, is called geochemical dispersion. Dispersion may be the effect of mechanical agencies or chemical processes. Purely mechanical processes of dispersion usually involve mixing but not differentiation of the dispersed materials into specialized fractions. In contrast, chemical and biochemical processes commonly create fractions of widely differing chemical composition. The more mobile fractions tend to leave their original host if adequate channel ways and chemical or physical gradients are available. Dispersion may be either deep-seated or surficial, according to the geochemical environment in which it occurs, and primary or secondary according to whether it occurs during the formation of the ore deposit or during a later stage.

Around magmatic and most hydrothermal deposits primary dispersion occurs in the deep-seated environment, and secondary dispersion in the surficial environment. In deep-seated dispersion the channel ways and sites of redeposition are generally the fissures and intergranular openings of deep-seated rocks. Surficial dispersion on the other hand takes place at or near the surface of the earth, where patterns are formed in the fissures and joints of near-surface rocks in the unconsolidated overburden, in streams, lakes, vegetation and even in the open air. Exploration geochemistry looks for traces of material that have been dispersed away from the ore body. These dispersion patterns may form over a considerably larger area than the ore itself, so that a correspondingly lower sample density is required for its discovery. In addition it is also necessary to know the pattern of elemental distribution in normal rocks so as to be able to distinguish the normal and the anomalous pattern related to a ore body.

In order to determine the existence and characteristics of different types of dispersion associated with mineralisation, 'Orientation Surveys' can be conducted (1) to define background and anomalous geochemical values, (2) to define adequate prospecting method utilizing the various available sample media and analytical techniques and (3) to identify the criteria and factors that influence dispersion (Fig. 3.4). Absence of anomaly in an area is also a significant observation while conducting geochemical survey. Since the basic objective is to prepare the geochemical maps which will guide the future user in right direction, attempts should be made not only to summarise the conditions where geochemistry can be used as a

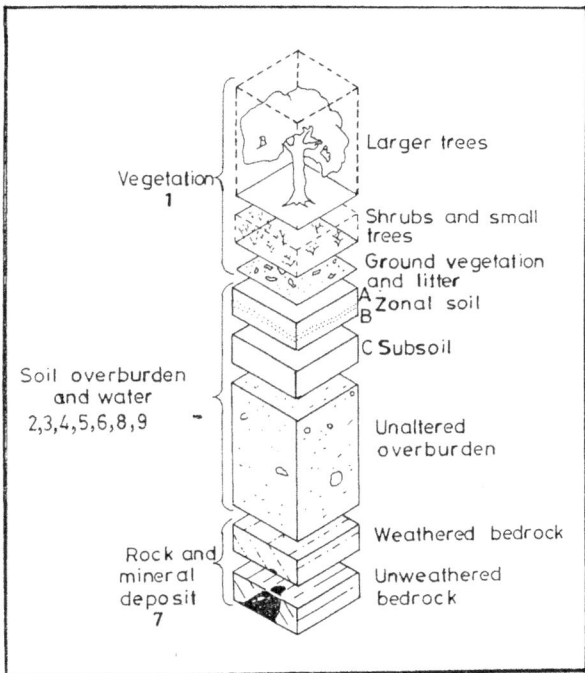

Fig. 3.4 The various environments and sampling material important in exploration geochemistry which are (1) plant, (2) soil, (3) bog, (4) water, (5) stream sediment, (6) overburden, (7) bedrock with mineral deposits, (8) pathfinder minerals, (9) boulder tracing etc. (After Levinsion, 1974)

reliable exploration tool, but also to identify areas where geochemistry can be reliably applied for mineral targeting etc. Orientation survey should be done before starting geochemical survey of any area irrespective of the sample media.

3.2.3 Types of Geochemical Anomaly

An anomaly is a deviation from the norm. An ore deposit being a relatively rare or abnormal phenomenon is itself a geochemical anomaly. Strictly speaking, an anomaly within a geochemical data set is defined as a chemical abnormality reflecting a disturbance of normal (background) chemical equilibrium caused by a mineralizing event. The objective of the interpretation of geochemical data is the recognition of such an abnormal concentration of an element in a sample associated with mineralisation compared to the concentration of the element in similar sample material from a geologically comparable but non-mineralised situation. But the presence of a mineral deposit in an area under investigation is not the only factor that can disturb chemical equilibrium. In fact, one of the greatest problem in the interpretation of geochemical survey data is to discriminate between "anomalies" that are due to concealed mineral deposits and those that are due to other causes. A simple and common case is the effect which a zone of slightly impeded drainage in an area of otherwise well drained soils may have on the contents in soils. In such case, there are two background situations: well-drained soils and poorly drained soils. Here exploration geochemist should recognize this fact and interpret the data accordingly.

Regardless of interpretative difficulties and deficiencies, the practical application of exploration geochemistry is based on the search for geochemical anomalies in samples of the earth's crust to assess the mineral potential of an area. It is necessary to consider the value of the measured parameter, the physico-chemical nature of the sample material, the physico-chemical environment from which the sample is taken, and the spatial distribution pattern of the measured variable. An anomaly, thus, is an abnormally high or low content of an element or element combination, or an abnormal spatial distribution of an element or element combination in a particular rock type in a particular environment as measured by a particular analytical technique.

In conventional interpretation, an anomaly cannot be recognized unless background is defined. It has been observed that the content of any element in any geological material cannot be represented as a single number. Apart from variations due to error in sample preparation and analysis that are inevitably present in the reported value for a particular sample, there is an inherent variability in the content of an element even in a small and petrologically homogenous granite or say, in B-horizon of residual soil samples overlying a homogenous rock. In statistical terms the recognition of an anomalous sample depends upon establishing the statistically probable range that occurs in the background samples i.e. the background population, and calculating (for some defined confidence level) an acceptable upper limit, called threshold of background fluctuation. Any sample that exceeds this threshold (lowest anomalous value) is regarded as possibly anomalous belonging to a separate population (vide Chapter 4, Fig. 4.1). Similarly, for negative anomalies threshold defines the lower limit of background fluctuation.

Anomaly may be of different types. Anomaly due to primary dispersion patterns is developed within the host rock/country rock during mineralisation. The mineralizing solutions or emanations, in case of endogenic deposits, pervade the country rock through micro- and macro-fissures or even through intergranular spaces of rocks, and depending upon the mobility of the elements the primary dispersion pattern (or halo) is formed. The theoretical shapes of dispersion patterns formed by diffusion and infiltration (leakage) are different. The 'metal content versus distance from ore body' curves are concave upwards in case of diffusion and more or less S-shaped and concave downward in case of infiltration or leakage (Rose et al., 1979). Leakage anomaly can extend hundreds of metres from the ore deposit, whereas diffusion anomalies are normally limited upto 30 m. Under favourable condition primary halos may extend 200-800 m above hydrothermal deposits controlled by steeply dipping planar structures of the host rock.

Anomaly manifested by secondary dispersion is a consequence of the dynamic processes of redistribution of elements in the zone of weathering. The pattern is an expression of the relative mobility of the elements in a dispersive environment. This may give rise to distinct anomaly patterns in residual overburden, transported overburden, natural waters or in drainage sediments in and around an ore deposit (either sub-cropping or concealed type). The secondary dispersion may have a symmetrical distribution pattern ('halo') around the mineral deposit under the normal condition, or a unidirectional distribution pattern ('fan') diverging away from the source or even a linear distribution pattern ('train') leading away from the source. Such fan and train develop when lateral migration of elements is effective due to weathering, glacial action or groundwater flow (Fig. 3.4). In geochemical prospecting for mineral deposits, residual soil is the most reliable medium for study of secondary dispersion pattern. Other media for goechemical studies include non-residual cover (transported over-burden), soil gas, stream sediments, water, vegetation (for biochemical prospecting and geobotanical prospecting) etc.

A review of the exploration practices followed in India indicates that sub-aerial weathering history plays an important role. Some work in this respect has already been carried out (Banerji, 1983). The weathering history of a major part of Indian Peninsula probably dates from Permian including broadly an early period of widespread lateritization followed by aridity. This gave rise to the formation of a deep weathering profile in many parts, which may subsequently have been partially or wholly eroded. Consequently, emphasis should be placed on the long weathering history and the influence it has had on element dispersion. The surface expression of mineralisation in a wide range of sample media like ironstone, gossans, weathered rock, soil, transported overburden, stream sediments, waters, plants and atmosphere, should be critically studied while sampling from Peninsular India. Such studies should try to deduce the geochemical characteristics of the sample media from multi-element data, wherever possible.

Butt and Smith (1980) derived a set of idealized models using data from case histories and published literature on Australia. They noted complete weathering profile, partly stripped profile or fresh bed rock in low relief, moderate relief and high relief areas. Under each geomorphic unit having high, low or moderate relief, there may be in situ, transported or no sample at all. Obviously there will be a number of such combinations and each one will require special attention from geochemical point of view.

3.3 WEATHERING AND RELATED LANDSCAPE

In any terrain the effects of weathering is most important as it decides the nature and type of mineral deposits that are likely to be generated from the country rocks in any geological setup. The products of chemical and physical weathering form a deep and extensive mantle over much of the landscape of Indian sub-continent including other continents of the world. An understanding of the nature and origin of this mantle is essential for (1) appreciating the constraints it applies to geochemical and geophysical exploration and (2) devising suitable and effective exploration technique for such terrain. The present nature of the weathering profiles, including their chemical characteristics, is a function of the combined effects of the various weathering episodes superimposed upon the characteristics determined by the parent lithology and topographic situation of the profile. The significance of these factors can be illustrated by the manner in which certain elements have been concentrated to form secondary deposits, or dispersed as primary mineralisation was subsequently weathered and eroded.

3.3.1 Profile Formation and Landscape Reduction

The weathering profiles are developed progressively, with each horizon formed from a progenitor, which resembles the horizon currently underlying it. The full sequence is thus not present until a considerable interval of time has passed. Assuming an initially high water-table, the effect of continued chemical weathering with little erosion of the insoluble products is the development of a *saprolite* from which mobile constituents have been removed but without deleting the original rock textures preserved by stable primary minerals and newly formed secondary minerals. Subsequently with the gradual dissolution of primary and secondary minerals and the growth of new structures, the original rock texture is destroyed. With time, settling and compaction of most resistant minerals take place and the concretions are developed giving rise to pisolitic horizons. Above the water-table, leaching under acid and oxidizing conditions may cause further reworking of the hydrous Fe oxides, so that the surface horizon may consist largely of residual sand composed mostly of quartz (except over the more basic rocks) and resistant accessory minerals like zircon, ilmenite, magnetite, chromite etc. Thus, when a full profile is developed, the fresh rock merges gradually (at times abruptly) with the leached saprolite, followed at the top a mottled zone with irregular Fe-oxide concretions. The mottled zone in turn grades to the ferruginous zone retaining rock texture at its base and increasingly pisolitic upward with an overlying sandy horizon accumulated largely as a residuum. The overall process is one of chemical wasting with continuing slow erosion resulting in an overall lowering of the landscape with time, as materials are eroded from the uplands and deposited in the valley.

The next stages in the evolution of the weathering mantle involve two steps, namely, (1) the gradual lowering of the water-table, which cause modifications to the profile (particularly to the upper horizons), and slow down the overall rate of weathering and (2) slope instability, giving increased erosion and sedimentation.

As the water-table declines to progressively lower levels, the horizons above it become exposed to highly oxidizing conditions through which there is a net over-flow of water. Dehydration of the upper-most ferruginous layer causes irreversible hardening and the formation of a duricrust. If due to climatic fluctuations or other causes, 'lateritic' weathering conditions become re-established with the water-table stable at a lower level, distinct zones of Fe-oxide accumulation may develop and get preserved. The net result is the profile commonly seen with the ferruginous and mottled zones overlying an extensively leached saprolite passing through a usually narrow zone of saprolitized rock into fresh rock. The resultant profile is the product of at least two phases of chemical evolution. Firstly, 'lateritizition' under high water table and secondly, 'leaching' under progressively lower water tables at times giving rise to typical gossan zones from the underlying sulphides.

3.3.2   *Chemical Effects of Weathering*

The chemical processes of profile development in the lateritic phase are chiefly:

(a) The loss of the mobile constituents such as the alkalis and alkaline earths, and subsequently of less mobile constituents such as combined silica and alumina.
(b) The retention of insoluble constituents as secondary minerals, such as kaolinite, other clays and Al-oxides.
(c) The mobilization of redox-controlled constituents, such as Fe and Mn, with reprecipitation as hydrous oxides near the water table.
(d) Retention and residual concentration of resistant minerals such as zircon, chromite, magnetite and quartz.

In the post-lateritic phase of lowering water-tables the processes are:

(a) Further leaching of upper horizons, especially in humid areas, with saprolite formation continuing at depth.
(b) In arid areas, precipitation of silica, carbonates and other components dissolved in ground waters as the hydrological regime became inadequate for their removal.
(c) The formation of duricrusts by the irreversible dehydration and hardening of ferruginous and siliceous horizons.

Profile variations related to bedrock lithology are also due to differences in the susceptibility to weathering of various minerals and mineral assemblages and in physical characteristics of the rock type such as porosity, joint etc.

The profile developed under given climatic conditions depends upon the drainage and hence its topographic situations. Different profile types thus form in different parts of the landscape. The processes of saprolite formation and residual accumulation described previously continue throughout, but have most relevance to the upland situation. On lower slopes and valleys the *in situ* weathering profile may have absolute enrichments of components derived from upslope – either in solution or in suspension. During periods of lateritic weathering, Fe-oxide and occasionally Mn-oxide enrichment develop in valleys.

Mechanical weathering, aeolian transport and precipitation of calcium carbonate or amorphous silica characterize profiles formed under an environment. The weathering crust where silt/sand is cemented by calcium carbonate precipitated from ground water, is known as *calcret*. Where the cement is amorphous silica, the crust is called *silcrete*. Silcretes normally develop during a phase of progressive aridity leading to alkaline groundwater. Some calcretes are also silicified. Both calcretes and silcretes grade into *playas* where high salinity and a strong chloride regime exist. Pedogenic calcretes are known to be abnormally enriched in uranium and gold, while silcretes also carry enrichment of REE and uranium besides Au, Ti, Zr, etc., depending on bedrock compositions.

In contrast, humid tropical weathering profiles broadly undulating to near-peneplained surfaces are dominated by chemical alteration and breakdown of the bedrocks under mildly acidic (carbonic/humic acids) water domain. Many elements including REE are mobilized to varying extents and either re-precipitated in the lower part of the profile, or absorbed by clay or iron and manganese oxides, or leached out of the system e.g. alkaline metals (Banerji, 1989). It is commonly believed that the end product of this process is a lateritic regolith, although some have emphasized that very good drainage is the only essential requirement for this purpose.

In the Indian scenario, three types of weathered profile are common, namely:

Type I: The ancient weathered profiles which remained exposed over part of the Indian Shield and which had possibly evolved at different times. They contain diverse and superposed chemical and mineralogical records of mechanical and chemical weathering and erosion (Banerji, 1983).

Type II: The ancient weathered profiles covered by shallow water sedimentary rocks or Deccan/Rajmahal Trap.

Type III: The weathered profile exhumed through differential uplift and headward erosion leaving only tell-tale marks of their earlier alterations.

Such weathered profile in the long run gives rise to placer deposits along coastal areas due to prolonged activities of sea water. These placer deposits are generally rich in REE and Rare Metals as in the case of east and west coasts of India. Such deposits are also common along the sea shore of different countries of the world.

3.3.3   Ore Formation during Weathering

The process of weathering gives rise to a number of important mineral deposits of different types, namely, (a) lateritic, (b) chemogenic and (c) detrital.

The deposits formed by lateritic processes include (i) bauxites, Ni laterites and certain types of Mn and Fe ores; (ii) residual deposits, Nb and rare earths over carbonates, Ti-V magnetites over certain gabbros; (iii) clay deposits including kaolinite, and (iv) supergene gold deposits etc.

In some cases, ore formation and upgrading is continued by post-lateritic processes. Thus, continued leaching of the upper horizons of the profile apparently led to further desilication of kaolinite and/or removal of Fe-oxide giving rise to bauxite. Initially, lateritic weathering in areas of subdued relief results in moderate enrichment of Ni (generally 1.0-1.57%) in ferrugenious saprolite. Upgrading occurs when uplift and dissection caused the water-table to fall and Ni released from the saprolite re-precipitated with Mg and Si derived from the weathering of primary silicates.

The most common chemogenic deposits (as present in Australia and some other countries) are the evaporites in arid environments. Concentration of dispersed components released during weathering may form such significant deposits. These include (1) the 'calcrete' type U deposits formed by the continued leaching of U, V and K from the predominantly granitic terrain and their evaporitic concentration and subsequent precipitation under suitable redox condition in major drainage area, and (2) uranium deposit concentrated in certain palaeo-channels in the sedimentary basins flanking the Precambrian craton.

Detrital or placer deposits may result due to sorting of resistant minerals during erosion, transport and sedimentation. Examples include concentration of Au, diamonds, cassiterite, etc. mainly along the coastal areas described earlier.

3.3.4   Implication of Weathering History

The influence that the various weathering and erosion processes have on the dispersion of the elements of interest has already been referred to above. These processes have obvious significance in exploration. Many of the characteristics of mineralisation become obscured and can be confused with other weathering products. The nature of the expression of a mineralisation in the weathering zones depends upon its landform situations during the history of weathering and this will determine the choice of sample medium and the analytical and interpretative procedures required. Oxidation and related phenomena of mineralized zones within country rocks produce oxides of base metal etc., which also produce gossanous outcrops that require special attention.

Most discoveries in countries like Canada, USA and Australia have been made in partly stripped or pediment zones where residuum is at or close to the surface. In such area, sampling of outcrop (fresh and weathered bedrock and ironstones), soils and in certain cases, stream sediments is carried out. But, there are regions where suitable areas for such sampling is not available, especially, if the area is partly or completely stripped of the profiles and is overlain by transported overburden. In such cases, sampling is to be done preferably by low cost shallow drilling.

In situations where a more or less complete profile is preserved, sampling of pisolites may be adequate, assuming these have formed in residuum (Smith et al., 1980). There are, however, many instances where the pisolites have developed in quite exotic transported overburden or have themselves been transported, and hence are unlikely to be a suitable sample media. In most instances sampling of the transported overburden itself is of little value. There are, however, exceptions. These include situations, where transportation has not been very great, such as in certain sand-plains, where heavy minerals or pisolites have remained fairly close to the source (Smith, ibid). A further exception is where hydromorphic processes have given a dispersion halo extending into the overburden. This is particularly common in valleys, where chemical weathering continued after deposition of alluvium – during or after the main lateritic phase.

Geochemical investigations in Peninsular India has by now established that the greatest obstacle in exploration of secondary dispersion weathered profiles is faced in tackling a laterite profile, which can completely mask the basement geochemistry at places. Similarly in the Himalayan tract of Extra-Peninsular India, transported profiles are very common. In such cases, it is necessary to obtain basement geochemical data from shallow boreholes, drilled at systematic intervals, to find out the primary geochemical dispersion with less cost. Similar types of problems are encountered in areas where a blanket of transported alluvium or desert-sand exceeds 10 m in thickness. In such cases also, one has to take resort to shallow drilling at systematic intervals to delineate the dispersion pattern of the primary values in the basement. In such cases vapour phase geochemistry may be tried cheaply, which will be discussed later.

In some of our geochemical exploration programmes it is observed that anomalies are recorded around old workings. Normally these anomalies are explained by contamination from old mining activities. In such a case, efforts should be made during the "Orientation Surveys" to measure the extent of contamination, or to find out whether sampling at appropriate depths could neutralize the role of contamination. This is a very important aspect, which is to be borne in mind, wherever one undertakes geochemical exploration in known prospects. It may be possible to define an optimum depth of sampling, which would ensure minimum contamination from old workings by this process.

3.3.5 Gossan and Its Role in Geochemical Exploration

Base metal sulphides occurring within the *in situ* rocks of any country when exposed to prolong weathering activity gives rise to typical Gossanous rocks that generally resembles ironstones. Such ironstones occur generally as surface or near-surface rock, in the form of gossans related to sulphide mineralization, and also as non-gossanous ironstones not related to sulphide mineralisation. Many economic mineral deposits of Indian Peninsula give rise to *in situ* gossanous outcrops. These include base-metal and other sulphide deposits of arid to semi-arid areas of Rajasthan, Gujarat, Madhya Pradesh, Andhra Pradesh, Karnataka etc. The large number of occurrences and excellent preservation of gossans in different physiographic provinces of India ensure that the gossans would provide an important medium for exploration (Talapatra, 1979). Gossans are the spectacular cellular outcropping mass of essentially limonite with gangue minerals over the sulphide ore bodies, formed due to oxidation and leaching of Fe-sulphides and other metallic sulphides that leave the hydrated iron oxide minerals behind often showing a zone of secondary enrichment (Fig. 3.5). Typical weathering features of underground sulphide lode zone are indicative of mineralization. Its bright yellow and red colour and weathering resistant nature make it a prominently visible guide for mineral exploration. Occasionally gossans may extend from a few tens of metres to more than hundred metres as observed in Dariba-Rajpura mineralised belt, Rajasthan. At times the gossan zones may itself contain sparse mineralisation that may be economical.

Gossan evaluation, including study of ironstone, in different parts of the world has so far provided a set of standard, sequential procedure for gossan recognition. This includes morphological description, selection of samples, study of chemical, textural and mineralogical parameters for outlining the processes of gossan formation and also to build up the historical development of gossan. The most important features of ironstones and their host rocks that should be noted in the field are: location, rainfall in the area, landform type, extent of weathering, regional geology with major rock types and important structural features, local geology with description

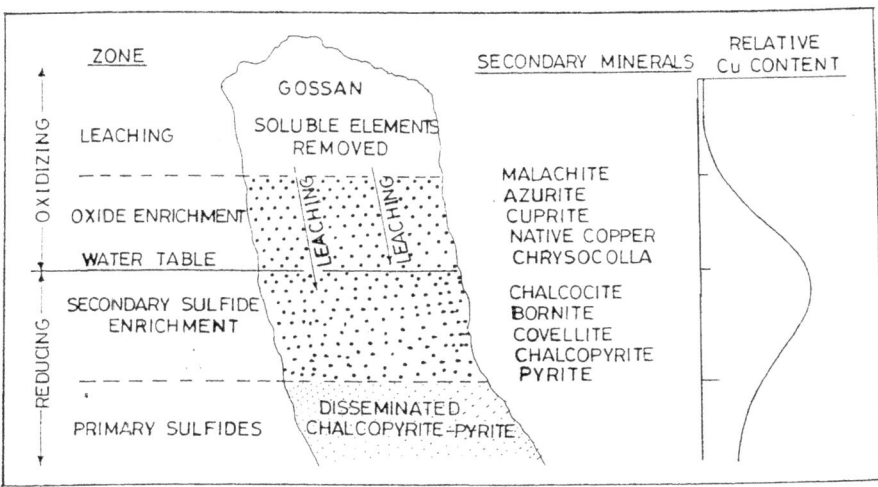

Fig. 3.5 Cross-section of a gossan zone and secondary enrichment above primary Cu-sulphide lode

of litho-types, alteration, microscopic textures and structure, nature of mineralisation including lustre, hardness, crystal form etc.

The most outstanding feature of the gossans developed from base-metal sulphides is that they have strongly anomalous target element (principally Cu/Pb/Zn) concentrations, for example, Cu in Khetri gossan and Pb-Zn in Dariba gossan, Rajasthan. The presence of target element in gossan outcrops as base-metal oxides, carbonates, sulphates and silicates, is a direct result of the arid climate and the presence of carbonate-rich wall rocks. Highly anomalous base-metal concentration occurring with exotic iron and manganese oxides in many deposits may be due to transport by upward moving groundwaters followed by precipitation near the surface in response to evaporation (Fig. 3.5).

Gossans developed over polymetallic deposits have, in addition to anomalous base metal contents, anomalous concentrations of one or more 'path finder elements' (vide Talapatra, 1979). Lithophile elements, barium and manganese also provide important components of gossans derived from stratiform Zn-Cu and stockwork Zn-Au deposits. A tentative list of target, path finder and lithophile elements expected in anomalous concentration in ironstones of various types (Taylor et al., 1980) is:

Volcanic massive sulphides	:	Cu, Pb, Zn, Ag,?Sb, ?Bi, Se, Hg, Zn, Mo.
Stratiform zinc-copper	:	Cu, Pb, Zn, Ag, Se, Hg, Ba, Mn.
Stockwork zinc-gold	:	Zn, Ag, Au, Hg, ?Sn, Ba.
Ultramafic nickel-copper	:	Cu, Ni, Co, Cr, (?Pt, ?Pd, ?Ir)
Fault ironstones	:	Cu, Zn, As
Laterites	:	Al, Si, Cr, V, Ti, Mn, P.

It is observed that multivariate discriminant analysis, using chemical analysis results (to define the discriminant functions), is successfully utilized to classify the different ironstones. If the data base containing multi-element analysis results of various ironstones is substantially increased, the classification of unknown ironstones of any area may be undertaken with confidence.

The technique of micro-texture analysis plays an important role in determining the source of anomalous target element concentrations in ironstones and this should be a routine procedure following chemical analysis for gossan evaluation. Although determination of textural characteristics under the microscope has identified many types of diagnostic textures for various types of sulphide to be identified from gossan study, their recognition is not always unequivocal. Determination of the mineralogy of the ironstones, particularly the identification of secondary minerals, facilitates the textural analysis and it also permits a preliminary understanding of the processes of surface enrichment of target elements. Different shades of colours like red, brown, yellow, coffee etc. sometimes help in identifying the original mineral though sometimes the colour mislead too as at Saladipura pyrite deposit, Rajasthan (Talapatra, 1979).

In exploration for polymetallic sulphide deposits in deeply-weathered terrain, the usefulness of multi-element geochemistry for gossan classification is now well established. The advantage of analyzing the chalcophile elements as a group, has been demonstrated by Taylor (1979) and others. In lateritic terrain of the nickel belt of Western Australia, R&D activity has shown that classification of Ni sulphide gossans is most successfully done by a multi-element approach using Pt, Pd, Ir and Te. Analysis of these may be adopted only in case of problem-solving purpose and not for routine analysis, as Instrumental Neutron Activation Analysis (INAA), required for this, is quite costly. Similar multi-element approaches using combinations of Ni-Cu-Zn-Mo-Mn and acid-soluble Cr has been very effective in distinguishing Ni-sulphide gossans from pseudo-gossans and other ironstones. Geochemical appraisal of the different types of gossans, specially those occurrences along the granite-greenstone belt of South India may indicate some anomalous areas where base metal sulphides including Ni and associated platinoid group of elements are likely to be present as significant deposits (Talapatra and Bose, 1978).

3.3.6 Geochemical Classification of Gossans of Rajasthan

Typical gossanous rocks after weathering of base metal sulphides are exposed along different states of Indian subcontinent out of which 10 different gossan occurrences of the three important mineralized belts of Rajasthan has been discussed here in details. The location map of the three mineralized belts with 10 different gossan occurrences of Rajasthan, India is shown in Fig. 3.6.

Many of the important base metal sulphide deposits, especially in the arid to semi-arid parts of the world, are characterized by well marked gossan zones occurring as weathered mantles above the unaltered sulphides. In India a number of such sulphide deposits are found, commonly polymetallic in nature, associated with typical gossan zones. A considerable number of such gossan occurrences have been located by the GSI in the western parts of the country, mostly in the state of Rajasthan. However, it has been observed that not all these gossan zones or zones of ferrugination are underlain by important sulphide mineral deposits with significant base-metal content (Talapatra, 1979). In this context, the evaluation of precise geochemical relationships between the sulphides and the corresponding gossan zones of the potential deposits appeared to be very useful in classifying the different types of gossans with special reference to their base metal content.

In order to establish the characteristics trace element association of the different gossans and their geochemical relations with the underlying sulphides, the following study was done (Talapatra, ibid). A suite of 210 samples of gossans, leached rocks and false gossans from ten different occurrences of three mineralized belt (Fig. 3.6), namely, (1) the Rajpura-Dariba-Bethunmi Zn-Pb-Cu belt (Dariba, Rajpura, Bethumni and Jasma), (2) the Khetri Cu belt (Khetri, Satkui, Dhanaota and Saladipura) and (3) the Alwar-Jaipur Cu belt (Kalighati and Naldeswar), together with 89 subsurface samples of sulphides from the mines and drill cores from four of

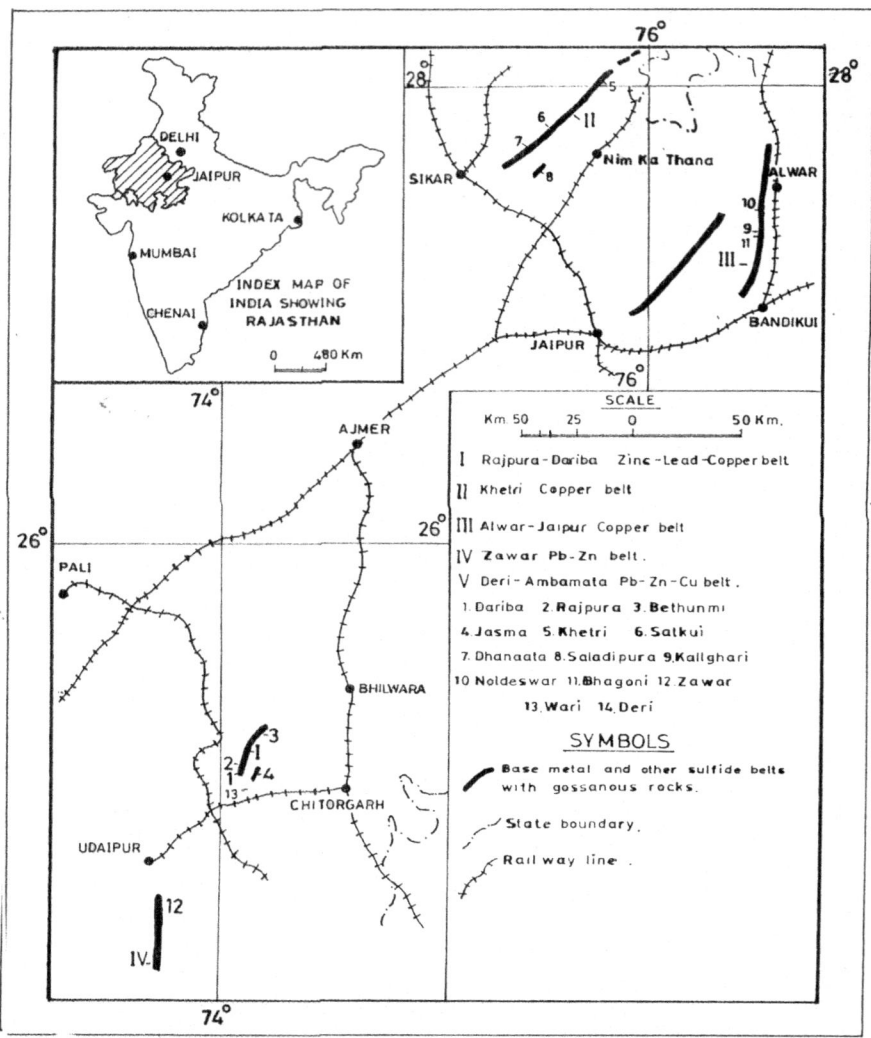

Fig. 3.6 Loation map of the three mineralized belts with 10 different gossanoccurrences of Rajasthan, India

the above occurrences were analysed for 20 elements, namely, Pb, Zn, Cu, Ni, Co, Cr, V, Sr, Ga, Ba, Mn, Sn, Mg, Sb, Cd, As, Ag, Mo, Bi and Zr. Statistical analyses of the geochemical data have revealed characteristics associations of elements within the individual gossan occurrences with some element pairs showing highly significant correlation coefficient values. Moreover, simple statistical parameters like arithmetic mean, geometric mean and standard deviation of the concentrations of some of these elements also appeared to be markedly different for different gossan occurrences. Thus the statistical studies of the geochemical data have been successfully used in these areas in conjunction with mineralogical characteristics for differentiating the "true gossans" from the "false gossans" and also for classifying broadly the different gossan types that have formed after specific base-metal sulphides along extensions of the known mineralized belt. Because of these results, detailed geochemical evaluation of the different gossan occurrences of Rajasthan promises to be highly successful in exploration for various types of sulphides. As such similar type of Gossans exposed in any continental area of the earth may be studied for locating the underlying base metal sulphides.

3.3.6.1 Description of Gossan Zones

An R&D project on geochemical and petrographic study of different types of gossans generally exposed on the surface along the underlying sulphide lodes of Rajasthan was taken up by the author from Geological Survey of India. In the following paragraphs a short petrographic and geochemical characteristics of this belt have been described (Talapatra, 1979).

Dariba-Rajpura-Bethumni Zn-Pb-Cu belt: The polymetallic sulphides of this belt are associated with high grade metamorphic rocks of the Pre-Aravalli Super Group (Deb and Pal, 2004) that form a part of the western limb of a regional syncline (Raja Rao et al., 1970). Spectacular, *in-situ*, hard, silicified gossan of various shades red, dark brown, black, yellow and white, with typical spongy boxwork is very well developed along a graphite schist-dolomite contact zone between Dariba and Rajpura villages (Plate 3.1) as a continuous body exposed in small detached ridges and hummocks upto Bethumni (Talapatra, 1979). Oxidation along this zone is upto 30 m thick in Dariba Block and reaches a maximum depth of 400 m in Rajpura 'B' Block. The Dariba-Rajpura gossan zone appears to be linked with the gossan zone of the area to the north through concealed extension zones in the Malikhera and Sindesar Blocks, where gossans are exposed only locally. The total strike length is thus about 17 km. Another small band of typical gossan about 600 m long is present within graphitic mica schist east of Dariba. Below these two gossan zones, two distinct polymetallic sulphide lode zones have been located by subsurface exploration and mining (Plate 3.2).

The lodes occur as steep easterly dipping concordant bands or lenses with some small shift due to local faults. Extensive old workings, mine dumps and slag heaps are found all along the gossan zones of the Dariba, Rajpura and Bethumni areas. Primary sulphide minerals present in these lode zones include sphalerite, galena, chalcopyrite, pyrite and pyrrhotite with some sulfosalts including geocronite, boulangerite, tetrahedrite tennantite etc. (Podder, 1972, 1974). The fourth occurrence of gossan in this area is in a chain of small hillocks trending north-northeast to south-southwest, lying to the west of Jasma village, about 7 km east of Dariba. The hillocks are composed of quartzitic rocks within garnetiferous mica schist country rock containing appreciable amounts of graphite. *In-situ* gossans having various colours including brown, black, brick-red, orange and yellow, with occasional angular boxworks, are exposed on the tops, as well as along the slopes of these hillocks.

Plate 3.1 Outcrop of multi-coloured gossan exposed on top of (top) Dariba hill and (bottom) Rajpura 'A' hill top with malachite stains (M)

Plate 3.2 Photographs of (top) mineralised load (L) zone (Zn-Pb) within dolomitic marble and (bottom) remobilized Pb-Zn ore within the host rock of underground mine level with galena (G) and sphalerite (S)

3D block model showing heavy mineral concentrations from core sample data of SK-105 Cruise off Gopalpur, Orissa

Block model showing heavy mineral concentrations from core sample data of Cruise SK-105 after removing desired/user defined segment

Khetri Cu belt: Base metal and other sulphides along this belt occur within metamorphosed rocks of the Delhi Super Group of Precambrian age comprising a thick pile of metasediments varying in composition from an arkose-orthoquartzite association having local bands of phyllites, carbon phyllites, schists, marble and amphibole quartzite, to a pelitic sequence of various types of schists, phyllites and calc-silicates. Host rocks in the Khetri area are generally garnetiferous quartz-chlorite schist, close to its footwall formation, namely feldspathic quartzite (Dasgupta, 1964, 1968, 1974). On the surface, rocks of the mineralized zones are characterized by ferruginous, gossan-like, leached rocks in shades varying from reddish brown, dark brown, coffee, yellow and orange to greenish stains of malachite. The sulphide load occurs in this belt as a number of detached lenticular bodies of different sizes with intervening lean and barren portions both along its strike and deep directions as inferred from subsurface data. The orebody consists mainly of pyrite, pyrrhotite and chalcopyrite with minor amounts of other sulphides including sphalerite, galena, arsenopyrite and cubanite. In the Satkui area, sulphides, mostly pyrite and chalcopyrite, occur along a suspected fault zone within similar types of metasediments with a characteristic dark coloured, hard gossan zone, occasionally steel grey, chocolate to dark yellowish brown, with abundant malachite stains and typical boxworks.

The mode of occurrence of sulphide in the Dhanaota area is similar to that of the Satkui, which is also characterized by dark coloured hard gossans with different types of boxwork texture. The pyrite pyrrohite deposit of the Saladipura area lies near the southwestern end of Khetri Cu belt detached from it within metamorphosed rocks of the Delhi Super group and is characterized by outcrops of typical gossan extending for a distance of about 7 km in two converging north-south trending zones, the eastern zone being 2.5 km and the western zone 4.5 km in length. The width of the gossan ranges from a few metres to as many as 120 m on the surface. The gossan is soft, porous and variegated, ranging in colour from coffee brown, maroon through brick red to yellow normally without any boxwork pattern. Sulphides, mostly pyrite, pyrrohotite and arsenopyrite with minor amounts of sphalerite, magnetite and rarely chalcopyrite occur within massive orebodies confined to amphibolites, schists, carbon phyllites and biotite-amphibole-quartzite. The sulpher content of the ore here ranges from 10 to 45 percent. All these areas were extensively drilled by GSI.

Alwar-Jaipur copper belt: Some scattered minor occurrences of base metal sulphides are reported along the belt, which extends broadly along two separate linear trends within predominantly arenous rocks of the Delhi Super group. A number of small copper deposits have been proved along this belt out of which only two occurrences were studied. Mineralisation occurs as sulphides, generally in quartz and calcite veins confined to grey phyllites intercalated with feldspathic quartzite which are at places intruded by basic sills. In the Kalighati area, a strongly limonitised band occurs within the small north-south trending ridge of brecciated ferruginous quartzite ('hornstone breccia') for about 3 km. Surface weathering of this rock has

given rise to dark brown, brick-red, orange and yellow colouration. Boxwork structures are rare. Subsurface exploration has not revealed any encouraging result; only
weak disseminations of pyrite and pyrrohtite in breccia and associated carbonaceous phyllites are noted. In the Naldeswar area, sulphide mineralisation appears to
be confined to an east-west striking fault zone within a massive ferruginous quartzite and marble sequence. Oxidation persists in this area to a depth 350 m, and only
some minute specks of pyrite and stains of malachite are found in the weathered
gossanised host rocks. Details of all these areas are available in unpublished reports
of GSI, Western Region.

3.3.6.2 Methodology

Detailed study of *in situ* gossanous rocks accompanied by thorough geochemical
analysis helps to locate base metal sulphide deposits along the mineralized belts. In
the following paragraphs the methodology of such investigations are outlined which
is valid for locating mineral deposits in any part of the world. In most of the mineralized belts described above, sulphide mineralisation is generally concealed by
post-mineralisation weathered cap rocks like typical gossans, leached rocks or false
gossans, formed by alteration of ferruginous host rocks above barren or lean portions without any significant base metal values. Unaltered primary sulphides were
detected on the surface outcrops very rarely. Hence in order to differentiate the
products of weathering of base metal sulphides from other ferruginous material for
exploration of mineral deposits, detailed geochemical appraisals of the *in-situ* gossanous rocks along the mineralized belts and their extension areas were carried out.
This was done by taking representative samples of the weathered cap rocks at regular intervals both along and across the strike of the mineralized zone. The samples
consisted of small chips taken from about 1 m² of each outcrop. The spacing of the
samples depended on the nature and dimensions of the gossanised rocks. Generally,
samples were at 10-20 m intervals across the gossan zones on traverses spaced 100
to 150 m apart along the strike of the zones. In places, samples were also taken at
much closer intervals. All the samples were powdered and passed through an
80-mesh sieve. Geochemical analyses of the samples were carried out by semi-
quantitative emission spectrographic methods and by atomic absorption spectrographic methods, which gave results for 20 elements, namely Pb, Zn, Cu, Ni, Co,
Cr, V, Sr, Ga, Ba, Mn, Sn, Mg, Sb, Cd, As, Ag, Mo, Bi and Zr. Chemical analysis of
gossan samples using modern analytical methods gives sufficient indications of
base-metal and associated economic mineral deposits, if any, underlying the gossanous outcrops that are exposed along the surface.

3.3.6.3 Discussion of the Results

Element distribution in gossans, leached rocks and the underlying sulphides of the Dariba, Rajpura, Khetri and Saladipura areas were studied with the statistical parameters like arithmetic mean (AM), geometric mean (GM), and standard deviation (SD) for the various elements. It appears from the distribution patterns and the statistical parameters of the base metal sulphides and other associated elements of the gossans of these areas reflect the assemblage of elements concentrated in their respective sulphides, and that are retained throughout oxidation in their weathered derivatives (Talapatra, 1979). For example, the Rajpura–Dariba mineralized belt is characterized by the presence of very high amounts of Cu, Ba, Mn, Mg and As in the gossan zone rocks (Fig. 3.7a and 3.7b), while their corresponding sulphides contain still higher concentrations of all these elements. Variations in base metal distribution in the gossan zones appear to be a function of both element mobility and specific differences in the environments of the several oxidizing sulphide bodies. The distribution of a group of elements is more helpful here than the use of single element or simple ratios. High values of Sb and Cd in addition to the above mentioned elements in the Dariba ore are due to the local enrichment of these elements within some of the sulphides and sulphosalts (Podder, 1974). Table 3.1 gives the matrix of correlation coefficients of the gossan samples of the Dariba block for 10 elements out of 46 samples. This shows that some element pairs have highly significant correlation marked by one or two stars.

The gossans of the Bethunmi area also show strikingly high values of a similar assemblage of elements, except Ba and Ag, which are concentrated less in comparison to the gossans of the Dariba and Rajpura areas. It appears that the entire belt from Dariba to Bethunmi via Malikhera, Sindesar Kalan and Khurd is likely to be mineralised. In contrast, the ferruginous gossanised rocks of the Jasma area do not contain the above mentioned assemblage of elements in comparable amounts, except for As. Thus, from the study of the element distributions and their interrelation, it may be possible to identify or characterize a gossan sample collected from extension areas of the known mineralized belts, and to decide whether it has formed due to alteration of the particular base-metal bearing sulphide lode or formed from some barren ferruginous rock units. Hence, such geochemical studies (Talapatra, 1979) may be used as a positive tool for guiding an exploration programme in this type of mineralized belt.

The element association of the Khetri Cu belt, however, exhibits a strikingly different assemblage in which only copper shows distinctly high values, and elements like Mn, Ni and Co show a moderately high range of values within the gossan zone rocks. The underlying sulphides corresponding to these gossan zone rocks also contain proportionately higher contents of the above mentioned elements. Here most of the elements within the leached gossan zone rocks are depleted, while Sr, Ba and Mn are enriched in moderately higher concentrations. Interrelations of some of the elements within the gossan zone rocks are displayed in the correlation matrix (vide Talapatra, 1979). Gossan occurrences of Satkui and Dhanaota, lying in the southwestern extension areas of the same belt, show almost identical element

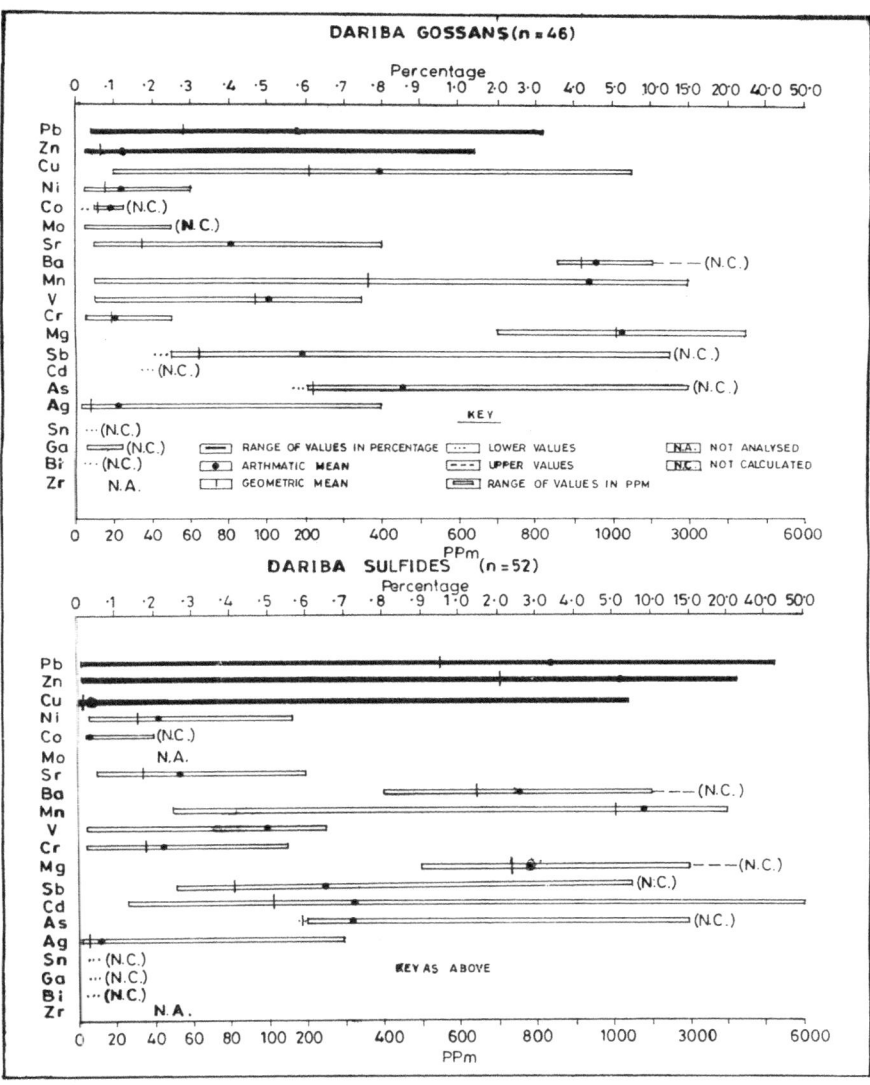

Fig. 3.7a Element distribution in the Dariba gossans and sulphides from mine levels and drill cores with some of their statistical parameters

associations having characteristically high concentration of Cu, Mn, Ni and Co. But the gossan zones corresponding to the sulphide deposits of Saladipura, which contain mainly pyrite and pyrrhotite, show different assemblages of elements having moderately high Zn, Pb, Ni, Mn and Zr. Here, the gossans of the east zone are markedly depleted in Zn, Mn and Ni, whereas the proportion of Pb, V, As and Zr is

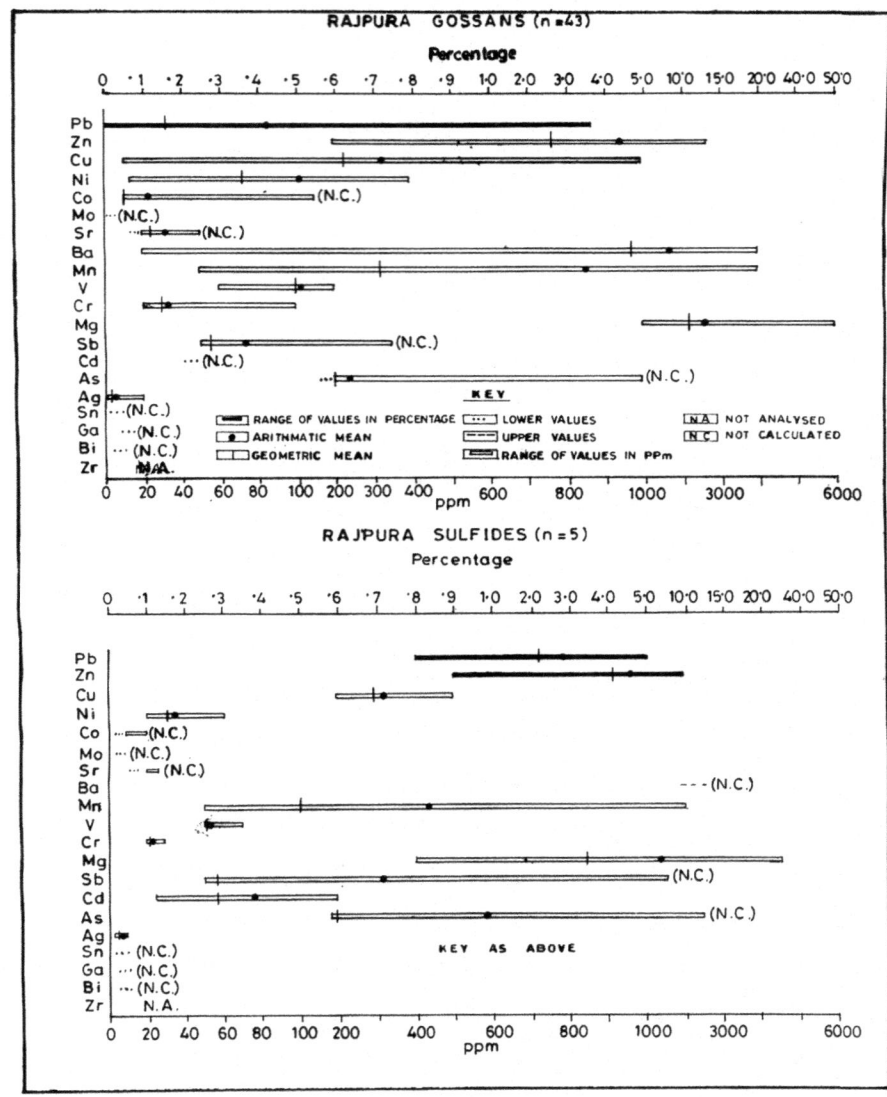

Fig. 3.7b Element distribution in the Rajpura gossans and sulphides from mine levels and drill cores with some of their statistical parameters

slightly enriched in the gossan zone rocks. Mutual relations of the different elements present within the gossans of the east zone also show a somewhat different pattern. The element distribution within the limonitised false gossan zone rocks of Kalighati and Naldeswar, like that of Jasma area, shows generally poor

Table 3.1 Martix of correlation coefficient of the gossan samples of Dariba block

	Pb	Zn	Cu	Ni	Sr	Mn	V	Cr	Mg	Ag
Pb	1.00	0.15	0.24	0.29	−0.21	0.17	0.01	−0.07	0.08	−0.29
Zn		1.00	0.18	−0.14	−0.19	0.11	−0.23	−0.11	−0.11	0.14
Cu			1.00	0.17	−0.45	0.26	−0.32	−0.27	0.08	0.08
Ni				1.00	−0.14	0.67	0.61	−0.08	0.60	−0.39
Sr		**			1.00	−0.03	0.24	−0.01	0.03	−0.09
Mn			**			1.00	0.38	0.11	0.92	−0.37
V			**				1.00	−0.08	0.48	−0.23
Cr								1.00	0.07	−0.05
Mg			**		**	**			1.00	−0.33
Ag			*		*				*	1.00

**Significant at 99% confidence level
*Significant at 95% confidence level

concentration of most of the elements; only Zn, Ni and Mn are present in moderately high amounts in both the Kalighati and Naldeswar occurrences, and Cu is comparatively enriched in the Naldeswar occurrence.

On the basis of the element association and the statistical parameters, the ten gossan occurrences so far studied can be tentatively classified into four distinct types as shown in Table 3.2. These are (1) the Dariba type (Zn-Pb association), (2) the Khetri type (Cu associate with minor Ni and Co), (3) the Saladipura type (Fe enriched generally with barren of base metals) and (4) the Kalighati type (Fe bearing, generally with minor amount of base metal sulphides). Refinement in exploration techniques for concealed sulphide deposits along the extension areas of known belts can be brought about by means of subsequent detailed geochemical investigation of the other gossan occurrences of Rajasthan and other parts of Indian continent on the lines of the work carried out in the areas described above.

3.3.6.4   Conclusion

Detailed geochemical study of ten different occurrences of gossan and gossan-like rocks from the three mineralized belts of Rajasthan shows that their characteristic element distribution and interrelations in conjunction with mineralogical characters help differentiate the "true gossans" from "false gossans" and classify these broadly into four different gossan types based on characteristic base-metal content of the underlying sulphides (vide Table 3.2). Many of the important base-metal sulphide deposits, especially in the arid to semi-arid parts of the world, are characterized by well marked gossan zones occurring as weathered mantles above the unaltered sulphides. In India we find a number of such sulphide deposits commonly polymetallic in nature, with typical gossan zones. It is also clear that the megascopic study of colour and texture do not always help to differentiate the type of gossans. Thus, with statistical analysis, various geochemical data from gossan zone rocks in parts of

Table 3.2 Geochemical characteristics of the different types of gossans

	Dariba type (Pb-Zn)		Khetri type (Cu)		Saladipura type (FeS)		Kalighati type (Barren)	
	Range	A.M.	Range	A.M.	Range	A.M.	Range	A.M.
Pb	0.04–3.15%	0.58%	5–100	14.29	20–2000	314.50	5-20	9.17
Zn	0.02%–1.44%	0.13%	200–400	357.14	200–4000	875.00	200-1000	533.33
Cu	20–1500	403.80	0.10%–0.88%	0.31%	20–300	64.90	30–600	162.50
Ni	5-60	27.28	10–200	33.21	15–100	28.40	10–200	78.33
Co	<10–25	5.54	10–200	55.36	10–50	10.90	5–50	21.67
Mo	<5–50	6.30	25–50	9.11	<5	<5	2.5–20	9.38
Sr	10–400	82.50	5–1000	113.57	N.A.	_	5–400	44.09
Ba	850–>2000	N.C.	5–300	118.93	N.A.	_	20–400	121.25
Mn	10–3000	936.63	30–2000	497.86	200–1800	385.90	30–1500	745.83
V	10–350	116.40	N.A.	–	50–150	81.90	N.A.	–
Cr	5–50	22.59	5–120	43.93	30–100	55.80	5–80	14.58
Mg	700–4500	1131.43	N.A.	–	N.A.	–	N.A.	–
Sb	<50–2500	200.00	–<100	<100	<30	<30	<100	<100
Cd	–<40	<40	–<100	<100	N.A.	–	<100	<100
As	<200–3000	455.43	–<1000	<1000	100–500	215.80	<1000	<1000
Ag	0.5–400	24.53	0.5–20	2.75	<1–8	1.77	0.5–2	0.71
Sn	–<10	–<10	<10–10	5.71	<20	<20	<10	<10
Ga	<5–25	5.61	2.5–10	6.25	10–25	13.20	2.5–10	4.38
Bi	<10	–<10	–<20	<20	<10	<10	<20	<20
Zr	N.A.	–	N.A.	–	30–250	71.70	N.A.	–

N.B.: All data in ppm except as otherwise noted, N.A. – not available, N.C. – not calculated

Rajasthan have proved to be good indicators of different types of sulphides. Such studies may also be of use in other mineralized areas having gossanous cap rocks in any arid to semi-arid parts of the world for exploration of underlying concealed basemetal sulphide deposits. This clearly indicates that detailed geochemical analysis from the *in situ* gossanous cap rocks from any part of mineral belt clearly suggests whether the gossan is 'true' or 'false'. So this type of geochemical study will help to locate concealed base metal and other related deposits in any part of the world from the study of gossans along the belt studied.

Chapter 4
DIFFERENT TYPES
OF GEOCHEMICAL EXPLORATIONS

4.1 INTRODUCTION

Sporadic geochemical analysis of samples were conducted by the Geological Survey of India (GSI) since its inception. But isolated geochemical surveys in known mineralized belts of India were initiated by GSI in the fifties of the last century and systematic geochemical work in GSI commenced only in the early seventies. However, the early attempts were sporadic in nature and the various geochemical surveys taken up by the regional offices lacked uniformity of approach. Invariably, there was time-lag between the field surveys and the receipt of analytical data which should be as minimum as possible for the interest of exploration work and future planning. Consequently proper processing and analysis of the data suffered. This is a problem common to many other developing countries of the world.

Classically, the geochemical survey programmes start with 'regional reconnoitory surveys' and gradually progress through 'orientation surveys' to detailed 'localized surveys' (Hawkes, 1972, 1976). In fact the Divisions of Regional Integrated Survey (DORIS) were set up in all the regions of GSI on this line and quite significant work was carried out till early 80's. Since then, it is increasingly being felt that the geochemical survey programmes in GSI are to be resumed on a national scale after appropriate reorientations which should take into cognizance the developing concepts of crustal evolution and metallogeny. An exercise in this direction has already been completed (Ray et al., 1987) with the preparation of a report on concept-oriented mineral exploration, where target areas for detailed exploration had been identified all over the country.

Tropical weathered profiles in land areas with cover of lateritic regolith are common in various parts of Indian Shield. Large deposits of bauxite and iron ore, moderate resources of manganese ore, concentrations of nickel oxide and of residual

© Capital Publishing Company, New Delhi, India 2020
A. K. Talapatra, *Geochemical Exploration and Modelling of Concealed Mineral Deposits*, https://doi.org/10.1007/978-3-030-48756-0_4

chromite have been established, but such cover in parts of East and South India can also be explored in detail for Platinum Group of Metals (PGM), Rare Earth Elements (REE) and gold concentrations, as also for signals of subjacent base metal sulphides (Banerji, 1989). This is also valid for other countries with Precambrian shield areas.

Application of geochemical exploration techniques is having limited scope in different offshore mineral survey and marine geology. The following areas/tracts may be geochemically explored: (1) Andaman-Nicobar Islands for Cu- Au- Mn and Pb, (2) Narcondum-Barren Island-Invisible Bank for Zn, Pb and Cu and (3) Active spreading centres of Andaman sea where Red Sea type Suphide oozes. In the first area there is a possibility of base metal suphide in association with 'ophiolitic' suite of rocks and cherts. Here Porphyry Cu–Au association in younger acid volcanics and Kuroko type situation in the spilitic lavas of the ultramafics of the islands are recorded. Around Narcondam and Nicobar Islands extensive indications of subarial erruptions have been noted where Kuroko-type mineralisation may be expected by geochemical methods.

4.2   METHODOLOGY

In order to carry out geochemical exploration in any parts of the world it is necessary to have a modern analytical laboratory which can analyze with precision all the different elements that are present within the field samples collected by the geologists from the area of study within a short time. It has been mentioned earlier that a geochemical map prepared from the analytical results of field samples can aim at projecting surficial chemical variations of different elements analyzed. Such a map with elemental contours helps in detecting the presence of any mineral deposits either exposed or concealed in the study area provided it is present there. Before undertaking a geochemical survey programme, its purpose should be clearly defined. Geochemical exploration data are collected at a variety of scales depending on the objective of the survey (Ginsberg, 1960; Hawkes and Webb, 1962; Elliot and Fletcher, 1974; Levinson, 1974; Reedman, 1979). In practice, geochemical surveys are of two main types: (i) regional scale and (ii) detailed local scale. In fact, all geochemical programmes should start with a regional scale survey in order to have some comprehensive idea about the regional variability of elemental distribution so that geochemical maps for individual elements are generated. This should be followed by local scale survey in selected areas based on the anomalous distribution pattern identified by regional surveys. Different stages of these two types of geochemical survey programme are as follows.

4.2.1 Regional Geochemical Survey

The main objective of this type of survey is to prepare regional geochemical maps on large scale ratios e.g. 1: 250,000 to 1: 2,000,000 or more. In this regard the technique of employing systematic stream sediment sampling data for compilation and generation of geochemical maps/atlas showing broad scale regional patterns has been widely accepted (Webb et al., 1978, Plant et al., 1988). This approach has also been followed in mineral exploration for a variety of metals in different parts of the world with considerable success. The underlying reason is that the products of rock weathering and soil erosion tend to be funneled down the surface drainage system. As a result, the active sediment in the stream bed approximates to a composite sample of those materials derived from the catchment area upstream from the sample site. Consequently, in normal circumstances, the pattern of metal distribution in the rocks and/or soils may be reflected to a degree in corresponding variations in the composition of the stream sediment. The geochemical data sets from stream sediment samples, thus, approximate to the average composition of catchment bed rock. This has been well reflected in a series of experimental surveys (Webb et al., 1964; Armour-Brown and Nichol, 1970; Nichol et al., 1966; Garret and Nichol, 1967) in which comparative studies of the relative merits of different types of media like rock, soil and stream sediment sampling were done.

These studies have demonstrated that the stream sediment samples should be the preferred medium for general purpose regional geochemical mapping, particularly in areas of heterogeneous geology and environmental conditions. The results showed conclusively that multi-element analysis of widely spaced stream sediment samples (at densities in the range of one sample per 2.5 to 180 sq. km. depending on local conditions) could be used not only to delineate potentially mineralized districts, but also to provide fundamental geochemical information relevant to the regional geology (Govett, 1983). In all such studies with fine-grained fraction of active mineral sediment, the elemental composition was found to be relatively stable under wide range of seasonal variations and topography. Multi-element regional geochemical maps prepared as an atlas from the stream sediment sample data are also found to be quite useful in the context of environmental problems in fields such as agriculture, pollution and public health, forestry, town planning and rural development, tourism etc. (Webb et al., 1964; Webb and Atkinson, 1965; Webb et al., 1968).

However, it should be remembered that the regional geochemical maps described above solve no problems beyond those of highlighting and assisting in the selection of areas wherein more detailed (and therefore more time consuming and costly) surveys may be conducted.

4.2.2   Local Geochemical Survey

Once the regional scale reconnaissance surveys (Webb and Howarth, 1979) are over, detailed geochemical studies may be taken up in selected areas based on the anomalous elemental distribution pattern in the regional geochemical map (Bhattacharya et al., 1984). Such detailed investigations are generally carried out in the follow-up stages of an exploration programme involving bed rock, weathered cap rock, soil, water and plant surveys, as is applicable in the local physiographical and geological conditions. Similar types of detailed geochemical work on local scale may also be simultaneously taken up along the extension area of known mineral belts/occurrences.

It has been shown that there are many variables in both primary and secondary environments that have profound effects on the dispersion of elements, the detection of which is the basis of exploration geochemistry (Levinson, 1974). As a result, every area in which exploration geochemistry is to be applied is likely to be different from a previously studied area in some way or the other. Therefore, a preliminary study called 'Orientation Survey' should be conducted in every new area. Thus, the normal practice is to undertake the 'Orientation Survey' first, (1) in order to select the appropriate sampling medium (in case of residual soil, to choose suitable horizon of the weathered profile of the area), (2) to standardize the sampling technique for the different media, (3) to optimize number of samples to be collected, (4) in the case of soil sample the size fraction to be analysed and (5) to define the criteria for selection of representative samples. During this survey, representative samples are collected from all the available media of the area under investigation. Especially in the case of soil samples, different horizons of soils are collected from some test pits reaching up to the weathered bed rock. After proper evaluation of the results of geochemical analysis of these samples, firm programme of the detailed survey is chalked out.

During this survey the areal extent of the geochemical anomalies are identified and plotted on maps on small scale ratios e.g. 1:2,000; 1:1,000 etc. Sampling is done on a grid pattern at regular interval. The work is usually accompanied by pitting and trenching whenever necessary for this geochemical survey to examine the nature and type of anomalies from the dispersion pattern of different path-finder elements and elements of interest.

4.2.3   Generation of Geochemical Maps

In this paragraph different types of geochemical maps that are generated from the analytical results of field samples have been described. In recent years GSI has started to collect geochemical samples for producing geochemical maps of study area and the maps showing elemental variation of each element with the contours are generated as a geochemical atlas of the study area. Various types of geochemical maps are generated from the different types of surveys which may be broadly grouped into two types, namely, (i) those which show the element concentrations at the sample locations (point-symbol maps, Talapatra et al., 1984) and (ii) those which emphasize the regional elemental distribution pattern at either local and/or regional scale. The second type may be thought of as reflecting the generalized (background) pattern of variation. Both the types of map normally display a single element for the sake of clarity and there are a wide variety of possible display methods (Webb et al., 1978; Govett, 1983). Such geochemical maps in the form of atlas of individual elements are found in most of the advanced countries of the world. In all such cases, good map design should be done by highlighting the information of primary interest, the geochemical variation, with necessary parameters on geographical, topographical or geological details. For this reason it is often convenient to overlay the geochemical maps on a geological map transparency. This has, however, become easy with the development of computer-based overlay techniques using GIS and other graphic packages. In fact, GSI has recently started detailed geochemical study of different types of samples collected from different parts of the country so that geochemical maps of different elements analysed can be generated in the form of atlas for the study area.

The regional maps showing distribution of individual elements may preferably be a point symbol map, smoothed grey level map using filters or colour combined principal component map. Other kinds of useful maps that may be generated by a plotter are: (1) filtered anomaly maps, in which the departure of data value from the general background trend is emphasized, (2) trend surface maps wherein regional compositional variation for individual elements are represented by contoured maps of the surface of best fit on the projected values of analytical results at each sample points and (3) multi-component maps, where attempts are made to show the simultaneous variation of a number of elements of interest (e.g., copper, lead and zinc), their ratios – simple (like Cu/Pb etc.), additive (like Cu+Pb/As+Ag) or multiplicative types (Cu $\times$ Pb/Zn $\times$ Ni), which often enhance the anomaly that are not prominent in the ordinary geochemical maps. Data may be plotted as profile curves also where it is desired to emphasize the distribution of metal/element or element of interest along certain cross sections or lines of samples. Multi element regional geochemical atlas prepared from stream sediment sampling are very useful for agriculture, pollution control, public health, town planning, rural development, tourism etc., besides selecting target areas for exploration of mineral deposits.

4.2.4   Anomaly Enhancement Techniques

One of the major trends in techniques of geochemical exploration is in the field of anomaly enhancement which is often applicable in blind deposits where anomalies in surface sample may be subtle or lacking. Some physical, chemical and statistical means are utilized for anomaly enhancement. Physical means include panned concentrate, magnetic concentration of heavy fractions and selected mineral separates like biotite, garnet, magnetite and hornblende. Chemical means employ dilute acid leaches, selective leaching of iron and manganese oxides and other cold extraction techniques. Statistical analyses for anomaly enhancement includes moving average analysis, trend surface analysis etc. In this regard mention may be made of simple ratios of some characteristic trace elements of the area of study, and their additive and multiplicative ratios referred earlier, which are being used for locating anomalous zones of mineralization by the geochemists of many developed countries. In all the cases mentioned above, insignificant or near background values of elements of interest are highlighted by these techniques specially in the case of concealed deposits.

4.2.5   Preparation of Samples

This is the most important part of a geochemical survey programme and should be conducted under the supervision of the experienced geologist. The following are the salient features that should be followed while preparing geochemical samples of different media.

Since rocks, ore and mineral samples will differ widely in their constitution, weathering features, size etc., the method employed in their preparation must be flexible and capable of adjustment to give good results for any type of sample. Initially the weathered surfaces of any such sample are removed, and sampling is done by chipping. Then the chips are split into small fragments after retaining a few representative chips for microscopic and other studies. These fragments are then crushed by mild steel crushers, followed by grinding in the pulverizing machine to minus 60 or minus 100 mesh, according to requirement. Once the sample is ready all the equipments like crushers, pulverizes etc. should be perfectly cleaned before preparing a fresh sample.

The preparation of soil and stream sediment sample materials depends on the type of analysis to be carried out on the samples. In certain cases complete analysis of the whole sample may be desired whereas in other cases only the minus 80 mesh fraction of the samples is required for analysis. The latter is generally suitable for most types of geochemical surveys using soils or stream sediments.

When the whole soil sample is required for analysis, same procedure as that given for rock sample is carried out, beginning from grinding step. Prior to this, all samples are required to be dried and all lumps of clay etc. broken up by means of baker's rolling pin or some other suitable non-contaminated instrument. Stone and other types of foreign material such as roots should be removed by hand picking or by a large mesh stainless steel sieve.

While only the minus 80 mesh fraction is required, the sample, after the lumps in it are broken up, is passed through 80 mesh stainless sieve. The under size is caught on a sheet of wrapping paper, rolled and retained as the sample. The oversize is discarded.

Orientation surveys, in the case of soil samples, should indicate the most appropriate size fraction required and samples of the regional geochemical survey should be prepared accordingly after selecting the appropriate soil horizon. However, if dry sieving is used, it may be difficult to separate clay aggregates from silt and sand fractions. In some soils and unconsolidated material, the clay fraction usually remains in the form of spherules, which may not disaggregate even during wet sieving.

It must be borne in mind that some amount of contamination will be involved for sample that need crushing prior to any fine milling. This should be kept down to a level that will not jeopardize the results of survey. Tests carried out in the Division for Mineralogy, CSIRO, Australia show that for the crusher with tungsten carbide, crushing surfaces introduce trace-elements like Bi, Cd, In, Mo or Sn was less than 1 ppm and Ag less than 0.1 ppm. This crusher is routinely used so that the low abundance levels of these elements in sample can be established. Tests show that crushing small quantities (say 10 gm) at a time by impact in hardened steel piston and cylinder gives the lowest general levels of trace metal contamination. This is a slow, manual operation and is therefore, applicable only to special studies.

The milling or grinding stage of samples, by its very nature, is prone to introduce contamination from the grinding surfaces. Tungsten carbide or hardened steel ring-grinders are quite adequate. But in case of highly abrasive samples such as silicified rocks are ground, levels of contamination may be quite appreciable. Tests conducted at CSIRO, Australia, have shown that upto 1250 ppm W and 120 ppm Co is added to samples of silcretes and quartzites on griding (Butt, 1981). In the grinding of gossan and ironstone samples by hardened steel swing mill grinders contamination is usually of little consequence because threshold levels in gossan are relatively high. However, in cases where thresholds are low, such as in multi-element geochemistry of laterite samples or stream sediment samples, milling in agate or alumina mortar is strongly recommended.

Laboratories carrying out trace-element analysis should be able to furnish in their report the analyses of the impact materials of their crushing apparatus. Analyses carried out on quartz blanks passed through the grinding or milling stage are not always satisfactory alternatives, but they provide complimentary information.

It is necessary in geochemical work to pay close attention to the prevention of cross contamination between successive samples during sample preparation. A practical way to remove this contamination, according to CSIRO scientists of Australia, is by light sand blasts using disposable quartz sand or glass beads, at times followed by immersion in an ultrasonic bath. Such additional care will certainly add cost to the sample preparation stage but increase greatly the reliability of the anomalies. Sample preparation is, thus, of paramount importance in all types of geochemical investigations and its follow up action, both for geochemical map preparation and for delineating ore deposits.

4.3 CONVENTIONAL TECHNIQUES
OF GEOCHEMICAL EXPLORATION

There are different types of conventional and non-conventional geochemical exploration techniques for unearthing mineral deposits in any terrain of the world. Let us first of all discuss the conventional techniques. There are many mineral deposits which are not exposed at the earth's surface. These may be covered by thick residual soil or may be buried beneath the rock formation. Some conventional geochemical exploration techniques are employed to locate these deposits, based on systematic measurement of one or more chemical property in different sample media like rock, soil, weathered profile, glacial debries, stream sediment, water, plants etc. (Hawkes, 1972, 1976). The chemical property most commonly measured is the content of the element of interest ('Key element') or 'pathfinder elements' which are likely to be associated with the mineral deposit under consideration (Taylor, 1979). The analytical result may detect the geochemical anomaly, if any, in the area that may ultimately guide to locate the mineral deposit (Chaffee, 1976).

The non-conventional geochemical exploration techniques are used in areas where the conventional geochemical exploration techniques are not applicable (such as areas with deeply buried deposits covered by transported soil, desert sand, talus material etc.). In such cases detection of soil gas, atmospheric gas etc. may be of much help. These non-conventional techniques are also capable of detecting the concealed deposits/lode zone in extension areas of known deposits covered by transported soil, desert sand etc. and will be discussed separately in the next chapter.

Geochemical exploration is a method of prospecting for mineral deposits, oil and natural gas by sampling and analyzing lithosphere, pedosphere, hydrosphere, biosphere and atmosphere (cf. Levinson, 1974; Reedman, 1979). The principle underlying this is based on the fact that mobile elements in mineral deposits have a tendency to attain various degrees of equilibrium distribution within the surrounding section, and that sampling of the easily accessible parts yields some suggestive information on the element distribution within the ore deposits lying at depth. The distribution or redistribution of chemical elements surrounding or associated with the ore bodies are a function of the physico-chemical condition of the local environment as has been mentioned earlier. The primary distribution pattern of elements is recorded in bed rocks, while secondary distribution pattern is noted in rest of the sample media like residual soil, weathered or gossanised rocks etc.(Fig. 4.1). Chemical and physical weathering brings about different types of secondary dispersion patterns of element. The conventional geochemical exploration techniques generally include sampling of bedrock, soil, transported overburden, stream sediment, water and plant. Before discussing these techniques, some idea on sampling methodology will be useful.

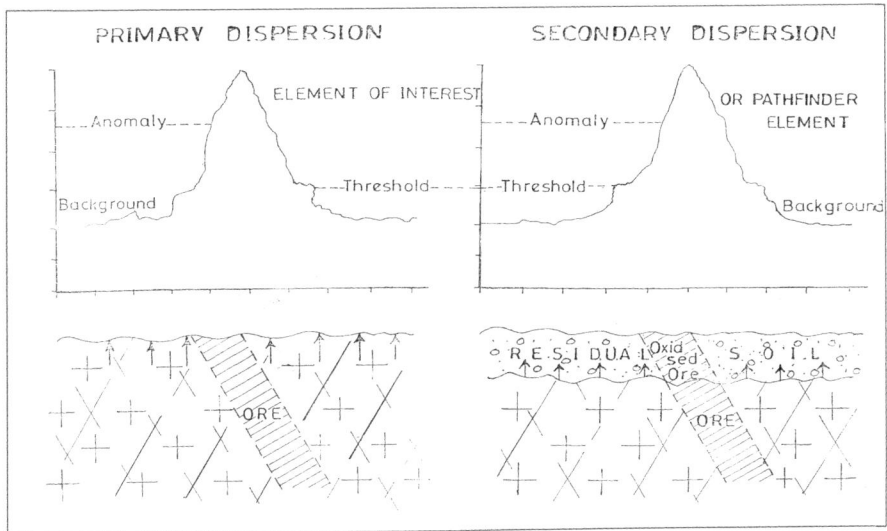

Fig. 4.1 Primary and secondary dispersion pattern of element along a section

4.3.1   Field Sampling Procedure

In general, geochemists undertake sampling programmes in order to achieve one of the two objectives, namely, (1) to describe the geochemical variability of an area, either statistically or spatially, (2) to carry out searches for mineral deposit-related targets. In the former instance the survey data are analysed by a variety of statistical methods. Sampling activity can be viewed as a data collection exercise in order that hypotheses stated, or inferred, by the geochemist may be tested. In the instance of search, or exploration, testing of the hypothesis would be that there are anomalies present in the sampled area. Whereas in the instance of a variability study the hypothesis might be that there is significant systematic regional variation present.

A concept fundamental to sampling methodology is that of target and sample populations. The target population is the totality of the rock unit, area, material type, etc. that the geochemist is interested in. It is the target population that is the ultimate subject of interest for hypothesis tests. In the vast majority of cases the target population cannot be sampled in totality because it would be logistically impossible; so only part of it is sampled. This collection of samples (to a geologist a sample is the individual and such terms as set and suite, which are used to describe the totality) forms the sample population and it is from this that statistical inferences may be drawn and estimates may be made; these may then be applied to the target population on the basis of subject matter reasoning. The choice of target and sample population is critical to good survey and research work. It is essential that the geochemist has a clear picture in mind of the hypothesis, which is to be tested, and the geological material to which the test is to be applied. Scale and size are also critical in sampling design. Not only the size of the mineral occurrence but more importantly the size of the geochemical halo, or dispersion pattern, around it. The boundary of the halo or pattern can be set at the upper limit of local (in a scale-dependent sense) background. Herein, a major advantage of geochemistry in exploration is utilized, i.e. the increase in search target size due to primary and/or secondary dispersion relative to the size of a mineral occurrence, or group of occurrences.

The collection and documentation of samples are also very important aspects of any geochemical programme, since the analytical data obtained from these samples are used for the subsequent decision making of the project. Depending upon the sample media, collection procedure also changes accordingly. It should be remembered that 'orientation survey' is required essentially to obtain the technical information for planning the geochemical survey, irrespective of any sampling media. In order to judge the level of precision of the analytical results in all the geochemical sample media, a check sample may be made after every tenth sample.

Next to collection and documentation is the preparation of the samples, which has been discussed in detail earlier. The essential geological parameters required for each sample may be noted as per standard formats (Banerji et al., 1982).

4.3.2   Litho-geochemical Sampling

It is the most common and widely used conventional technique of geochemical survey, which is also called bedrock sampling (Chaffee, 1976). In this case fresh chips from about 10 different points of an in situ outcrop are taken. The outcrop should be more or less homogeneous in composition. The entire chips are crushed in iron and porcelain mortar to pass through −100 mesh ASTM keeping aside some chips for microscopic and other studies. After two intermediate stage of coning and quartering, the powder is sealed in numbered alkathene bags weighing approximately 200 gms. in two bags (for keeping one set of duplicate sample). Similarly, in situ weathered/leached rock, where fresh rock is absent, is sampled for finding out the characteristic trace-element assemblage. Once the results of trace element analysis of the sample media of an area are available, these can be diagrammatically presented in different styles; one such method of presentation is shown in Fig. 4.2 which presents the contents of the elements – the average and range of the contents.

In an area where rock exposures are present, background samples of typical rocks from unmineralised portion of the area must be collected for comparison with the anomalous samples. The details of the orientation survey will vary with the type of ore being sought, the scale of the survey, stage of exploration, the type of material that can be sampled etc. Hence the type and amount of sample and the method of sample collection should be standardized during the orientation survey. Indicator/ pathfinder elements are usually relatively homogeneously distributed in primary dispersion patterns, in such case relatively small specimens of rocks may be adequate. But, in contrast, leakage anomalies are commonly localized along fracture, and the metal values are very erratically distributed out in the scale of

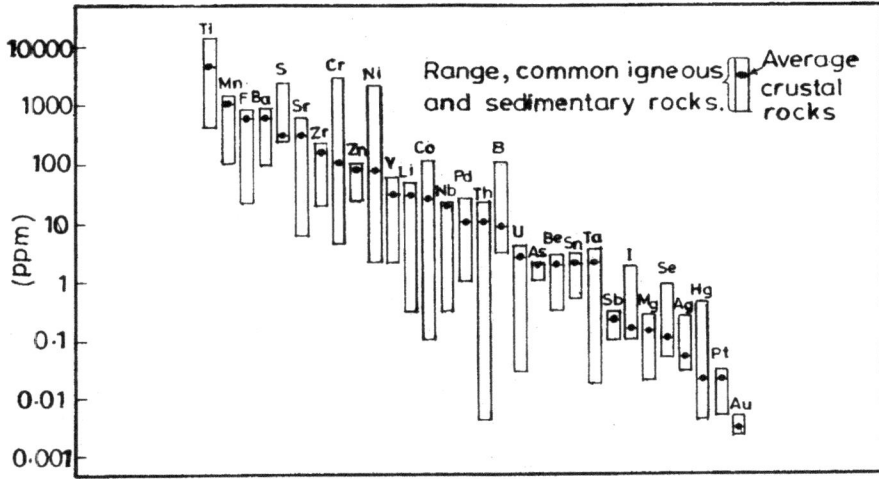

Fig. 4.2 Trace element data presentation in average crustal rocks showing the normal range and average

hand-specimens. In such cases, analyses of an aggregate of small chips from randomly chosen sites over several square metres of outcrops is generally preferable to analyses of a single piece of rock. A few chips of representative rock unit should also be collected for petrological studies under the microscope.

It is essential that the results of orientation survey for leakage dispersion should be interpreted in terms of the geology in three dimensions and care to discriminate wherever possible between primary dispersion and secondary redistribution of metals by percolating meteoric water (Fig. 4.1). In the later case it is very helpful to determine the primary ratio of metals possessing markedly different mobilisation in the zone of weathering. Thus, in supergene redistribution patterns, the ratio of zinc (mobile) to Pb (immobile) would be expected to show a marked increase compared to Zn/Pb ratio in primary ore.

The procedures for selecting sampling patterns and for locating and identifying samples are very important in litho-geochemical surveys. The sample spacing for detailed survey seeking veins or small ore bodies should be close spaced so that at least three to six samples fall within an anomaly and define a clear trend towards the ore. In detailed surveys a regular grid or spacing along traverses is normally preferred. Since large areas of complete outcrop are rare, rock samples cannot usually be collected on a rigid grid or rigid spacing along traverses. If exposures are relatively abundant, the closest outcrop to the grid point may be sampled. Alternatively, the area or pluton to be sampled may be divided into block/grids and one or more samples are selected from each block. It is important that sample be selected in as unbiased a fashion as possible so that they are representative of the area rather than units especially resistant to weathering or preferentially exposed along stream.

Considerable care must be taken in the preparation of rock samples in order to avoid contamination. Samples from underground workings and outcrops may be coated with debris from blasting, smeltar fumes, or staines and films from migrating surface or ground waters. Before grinding the samples, all such artificial contaminations should be removed by breaking or diamond-sawing off the effected portions. The quantity of samples varies from 500 gm for fine grained rocks to 2 kgs for very coarse grained rocks.

Leakage anomalies produced by hydrothermal process are strongly controlled by structure and permeability of the host rock, so that ore is not necessarily located directly beneath a surface anomaly. Careful geological mapping followed by tracing of the anomalies in drill cores may be necessary to determine the location of blind ore at depth. Distinction between supra ore anomalies indicative of blind ore bodies, anomalies around exposed one and sub-ore anomalies in the roots of eroded ore bodies is an important issue. For many types of hydrothermal ore deposits the zonal relations of elements may be used to make this distinction. For example, if Cu ore is being sought the presence of anomalies in Co, Mo and Cu at the surface would suggest that the samples were in or near the roots of eroded ore, and that if good ore is not present at the surface, then none is likely in depth. Orientation studies on known ore bodies within the district are very much desirable to confirm that the general zoning patterns are applicable in the district in question or not.

4.3.3 Weathered Bed Rock/Gossan Sampling

This includes rocks that have been chemically altered, have original structural elements and are essentially *in situ* with little or no lateral physical movement. The fossil (paleo) profiles may be completely or partly truncated, and either exposed or obscured by residual soils or transported overburden. Structural features preserved in the weathered rock may include bedding/schistosity, vein or lithological boundaries. Fine textural features expressed by the arrangement of the primary mineral constituents of the rock may be preserved, and in such case the weathered bedrock is referred to as 'saprolite'. The top of residual profiles is often marked by a stone-line comprised of the debris of resistant quartz veins smeared out at the base of creeping mass of soil or colluvium. Such a stone line marks the transported/residual interface. In poorly dissected areas weathered bedrock is the only readily accessible sampling media (Talapatra, 1979).

Weathered bedrock outcrops can be chip-sampled as described in the case of bedrock. But it is liable to bias, since only the most resistant/secondarily hardened rocks are exposed which may not be proper representative of the unexposed rocks. Pitting and trenching is useful in detailed exploration where bedrock is relatively soft, gossanous, poorly developed or where there is a thin cover of soil or transported overburden. It enables a more accurate assessment of bed rock and its relationship, permitting accurate channel sampling. Intermittently outcropping or concealed weathered bedrock is best sampled by drilling – such as auger, rotary air blast, rotary percussion or diamond drilling. The samples collected are usually composites of cutting at 1 or 2 m intervals. Additional samples may be collected on contacts or where apparently gossanous fragments are present. Samples are frequently subject to contamination, particularly where clay rich samples are taken from near the water table. For the correct selection of analytical samples and correct data interpretation accurate logging of drill cuttings are essential.

Besides these, all the methodology and precautions taken in the fresh rock sampling will also be applicable in altered *in situ* rock masses or cappings.

4.3.4 Pedogeochemical/Soil Sampling

Soil is the unconsolidated weathering product that has developed virtually *in situ* on the parent material from which it is derived. It is formed as a result of interaction of climate, organisms, topography and drainage over a prolonged period of time. The parent material may be either fresh or weathered bed-rock (including mineralisation) or some form of transported overburden. Usually the samples developed on residuam are of significance as an exploration media. Care must be taken so that samples are collected from an equivalent horizon throughout the area of investigation. Differentiation of soil is less significant in arid areas and here soils represent more directly their immediate parent materials. Nevertheless, carbonates and sulphates may accumulate in some soil, which may either act as diluents, or cause precipitation and concentration of metals such as copper etc. since differentiation can cause element partitioning where careful documentation is important.

Sampling is done from small pits, generally 70-80 cm from the surface, laid in a grid pattern from the 'B' horizon (where this horizon is prominently present in the study area). The soil is sun dried and passed through 100 mesh ASTM after coning and quartering. Two samples weighing about 100 gm each are prepared from the soil samples. Where the nature of overburden and basement does not change over a considerable stretch, one composite sample made from 3 to 4 successive soil samples, as the case may be, is sufficient. In such a case the composite sample is prepared by mixing equal volumes of all these soil samples and then reducing the volume by coning and quartering. This would reduce the volume of analytical work without diminishing the efficiency of sampling.

Soils can be sampled by using hand or power auger. Care must be exercised to sample a consistent soil horizon as mentioned before. A constant sample depth is, however, not recommended as the nature of the profile can change markedly over short distances. The 'C' horizon soil is often selected since it is the most consistent horizon. It is important to note here that the soil sampling should be based on prior detailed orientation survey mentioned earlier.

A completely thorough orientation survey for the soil sampling in a new area starts with the collection of a series of vertical sections through the soil profile, arranged as a traverse across the sub-outcrop of the mineralized ground. Comparable profiles from background areas on either side of the deposit should be sampled at the same time. Preliminary analysis of the −80 mesh to 100 mesh fraction for the predominant ore metal or associated pathfinder elements will usually suffice to show whether an anomaly is present or not. Analytical determination should include the total metal content (Me) and, in case of relatively mobile elements, the readily extractible metal content (CxMe) by one or more methods. The choice of extractants is based on the possible modes of occurrence of the metals in question. Selected anomalous and background samples should then be subjected to a series of experiments to determine the range of concentration of the key elements, as well as the size fraction and analytical method that shows the greatest contrast between anomaly and background. When dealing with elements, that characteristically occur in the soil as components of readily identifiable clastic minerals, the heavy mineral fraction of the soil should be examined mineralogically. On the basis of these experiments, proper procedures of preparation and analyses that show the maximum effective contrast of anomaly over background are selected. Further pitting may be necessary to cover fully the width of the anomalous pattern. Analysis of the complete suite of samples by the selected chemical procedures provides the basic information for choosing the most practicable horizon for sampling i.e. the minimum depth at which an adequate anomaly contrast is obtained over the greatest width. Replicate samples at selected points may be collected to determine sampling error.

While the foregoing procedure is generally applicable to most soil surveys in a new area, each problem will usually call for some special modifications and additions in the design of the orientation survey. Pitting may not always be practical, or even necessary, if adequate information is available from a closely comparable area. In such cases, the orientation study may be restricted to determining the metal content only of the near-surface soil horizon, preferably including the "B" horizon. Figure 4.3 (a & b) gives a 3-D sketch of soil and stream sediment anomalies of a particular mineralized area.

4.3.5 Transported Overburden/Talus Study

This is material of exotic or redistributed origin, such as alluvium, colluvium, sheet-wash, sand (fluvial/desertic), gravel, aeolian, clay, piedmont fan deposit and glacial debris that blankets fresh or weathered bed rock (Cameron, 1977). Some deposits are wholly or partially consolidated, with a variety of cementing media. They may include materials from the erosion of earlier lateritic profile, such as lateritic gravels, pisolitic laterites, silts and clays. The overburden may form thick deposits, exceeding 100 m. as in places of West Australia, North-East India etc. Normally, the distance of transport and degree of mixing and dilution eliminates the possibility of locating a mechanical dispersion halo from mineralisation.

Transported overburden is not usually sampled. Instead of that, systematic grid drilling to weathered/fresh bedrock below is preferred. This approach is costly since drill spacing have to be close, as the chemical dispersion in the bedrock is generally limited. Seepage anomalies and biogenic concentrations are perfectly useful in

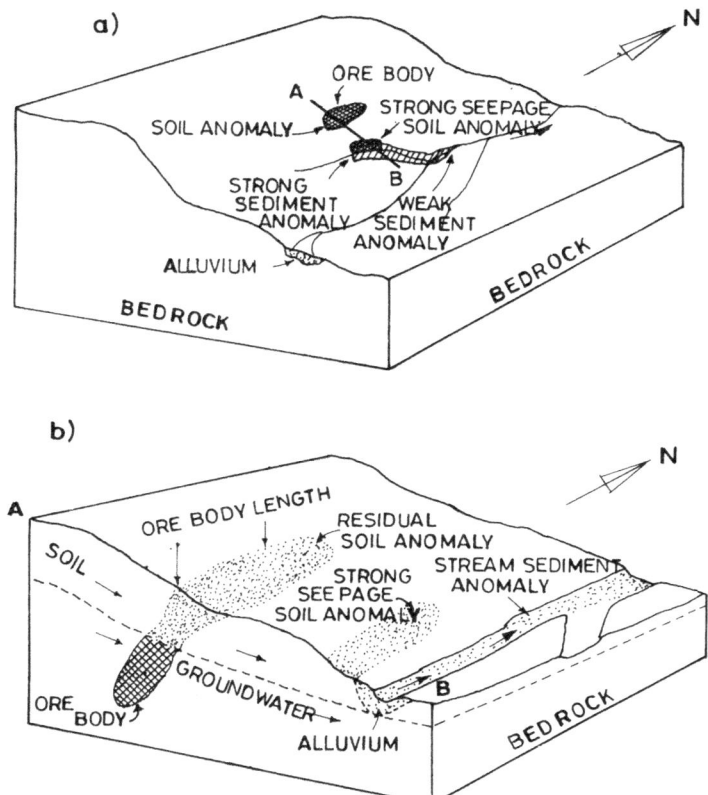

Fig. 4.3 (a) Block diagram showing the general patterns of stream sediment and soil anomalies in areas with essentially residual soil. (b) Diagram based on section A-B of figure (a)

some areas because of the ease of sample collection. If deep sampling is contemplated drill whole sampling should be tested to determine the time and effort to sample typical profiles to various depths.

Organic overburden (bogs, peats etc.) should, if possible, be test sampled in both anomalous and background areas in the same way as described above for residual soils. The distribution of metals at different depths in the profile and the (CxMe/Me) ratios of mobile elements should be determined. Careful attention should be directed to possible relations of metal content in the organic samples to the chemical character of water entering the bog (Eh and pH).

The best way to locate any concealed base-metal sulphide deposit or lode zone is to apply some of the non-conventional techniques based on vapour-phase geochemistry described in the next chapter, specially where cover material are transported soil, alluvium, wind borne desertic sand, talus material etc. The selection of sample pattern is determined primarily by the size and shape of the target based on the regional geological set up and, if possible, from the available geophysical data of the area of study. In looking for superjacent anomalies, the most suitable pattern is a simple rectilinear grid of samples of trapped soil-gas taken at equal intervals on the ground along lines at right angles to the elongation direction of the concealed lode zone, provided a steady de-gassing process takes place through the porous transported overburden of the area. Care should be taken so that at least two or more traverse lines intersect the anomaly of interest. This means that the traverse-line interval should be not more than one-third of the minimum economic strike length. Sample points along the line should then be spaced at intervals that ensure that at least two points fall within every important anomaly. Thus the interval is fixed by the probable minimum width of the expected anomalies.

4.3.6 Stream Sediment Sampling

It involves concurrent analysis of coarse and fine fraction of the active stream sediments as well as a continuous mineralogical scanning of the heavy fractions in the field with the aid of binocular microscope. Active stream sediments are the most representative sample type since these comprise composite samples of weathering products and have a limited Eh range as these are in contact with atmospheric oxygen (Garrels and Christ, 1965). Such samples have been used for geochemical survey in areas ranging from tropical rain forest, through savannah and desert, to high mountain ranges and arctic regions (Plant et al., 1988). Where stream sediments contain an important detrital component (for example, in higher latitudes) sieving should be employed to ensure that adequate quantities of the appropriate size fraction (generally, −80 to 100 mesh) are collected. Stream sediment sampling is generally confined to different order of tributaries. The prospecting technique normally includes systematic sampling of the entire catchment area of the tributary drainage system at all confluences and at regular pre-determined distances upstream, followed by analysis for the ore metals and associated elements, with a view to detecting anomalous dispersion trails, if any.

In order to minimize the time and cost of sampling, sometime tributary-road intersections are selected for such sampling (Webb et al., 1978) since the objective of the regional survey of stream sediments is to detect broad scale geochemical pattern only. However, it is a normal practice to carry out a preliminary orientation study to establish optimum techniques for the particular problem or terrain to be surveyed with the judicious combinations of sampling, analytical and data processing techniques. Generally at each point of confluence, three samples, one from the minor feeder channel, one downstream from its point of confluence and third one from upstream of the confluence point are taken for finding out the geochemical characteristics of the catchment area. In this connection mention may be made of the discovery of Balda and Dewa-ka-Bera tungsten occurrences of Rajasthan from stream sediment survey by G.S.I. during late 1980's. Panned concentrate samples from active stream sediments, in case of gold exploration, generally give better result (cf. Bhattacharya et al., 1984). In regional geochemical surveys, sample densities vary considerably. It has been observed that 60% of the surveys conducted in different countries of the world used densities greater than one sample per 5 km^2 (Plant et al., 1988). Stream sediment sampling is replaced by lake sediment sampling where the surface drainage is dominated by lakes.

Let us work out in detail the sampling of stream sediments. The topographic base map for these are generally 1:100,000 or 1:50,000 scale maps for regional geochemical survey from which large scale geochemical maps of the country may be prepared on 1:250,000 to 1:2,000,000 scale. Sampling traverses are laid out upstream along river courses and the most characteristic sampling points are located as shown in Fig. 4.3 where reconnaissance exploration followed by detailed studies using soil samples of the anomalous area is shown.

According to the system of notation adopted by geographers, river/stream having no tributaries are called first-order rivers, confluence of two first-order rivers yields a second-order river, and two second-order rivers form a third-order and so on. Rivers of higher orders can receive any number of lower-order tributaries without a change in their order. Traverses usually begin at the mouths of fourth order, sometimes fifth-order rivers and terminate at the mouths of second-order rivers (Fig. 4.4). Ordinary samples along the traverses that follow river courses are taken at 500 m intervals, while higher order rivers are not sampled.

In first-order rivers directly joining higher-order ones, at a distance of 100 to 200 m upstream from their mouths and at the end points of the traverses, two

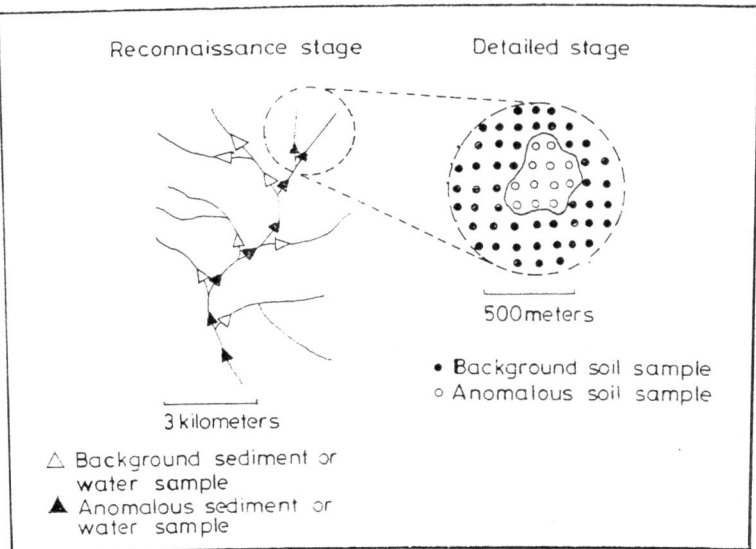

Fig. 4.4 Reconnaissance exploration using stream sediment or water, as contrasted to detailed soil sampling in anomalous catchment area of the stream

samples are taken spaced at 20-30 m. This makes the surveys more reliable and less laborious because the traverses are shorter.

In general a sample consists of sandy-clayey alluvium from a dry flood plain or bed of a stream. The initial weight of the sample may be, say, 150 to 200 gm, which is likely to yield some 50 gm of minus 1.0 mm fraction. Samples are collected into pre-numbered little bags or special 10 × 20 cm paper packs. Records of sampling in a field notebook are accompanied by those of geological observations and by drawing an outline sketch of the simplest type, which will provide identification of traverse points on the basis of local landmarks. However, precise location of the sample in terms of latitude and longitude may be made if a hand held GPS is used during sampling.

The field treatment of samples consists in their drying, minus 1.0 mm screening (with the removal of coarse fraction), and taking into a numbered paper envelope, after which the samples are sent to the laboratory. Normally for locating gold or any other precious metal a river basin is chosen where there is some earlier report. Bhattacharya et al. (1984) collected 18 stream sediment concentrate samples from streams of different orders of the Sonapet valley area (with catchment area of approx. 135 sq. km. of Singbhum dist., Jharkhand) which indicated 0.1 to 5.6 ppm Au. From these results three blocks have been demarcated (Figs. 4.5 and 4.6), where different types of vein-quartz traverse the country rocks, and fine specks of gold from channel sands are recovered by the local people by panning. Although the quartz veins of the country rocks are likely carriers of native gold, the exact source of this alluvial placer gold trapped in the channel sands, could not be identified. Further detailed geochemical analysis of the bedrock samples of the three blocks shown in Fig. 4.5 were suggested for locating the deposit, if any. It is concluded that the stream sediment concentrate samples are suitable for identifying the target areas, which can subsequently be explored by detailed bedrock sampling along with some sub-surface sampling for locating the mineralized zone, if any.

The modern procedures of conducting stream-sediment surveys are based on extensive computerized data processing for storage, treatment and production of element-wise geochemical maps using suitable GIS software. This, of course, does not rule out a direct participation of a geochemist in data treatment, primarily during the geological and quantitative interpretation of the detected anomalies and appraisal of ore element dispersion pattern. There are different modes of representation of trace element values from stream sediment samples. One such example has been shown in Fig. 4.7 which indicates the likely location of fertile pegmatite with beryl mineralisation in an area.

4.3.7   Hydrogeochemical Sampling

The samples of water from springs, wells and seepages on analysis give indication of mineralisation, if any. One-litre polythene bottle with screw cap are used for single element analysis, and three litre bottles for multi-element analysis for this

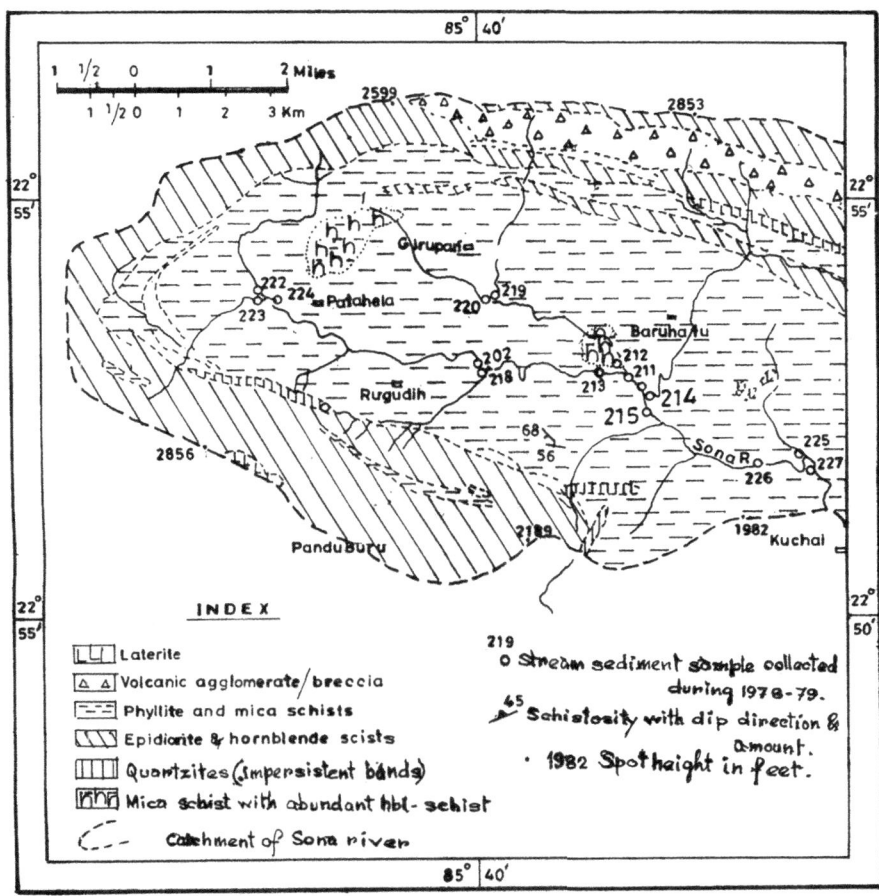

Fig. 4.5 Geological map of the Sonapet valley area, Ranchi district, Jharkhand showing the location of geochemical samples

type of sampling are found to be ideal. Highly sensitive and precise analytical instrument should be used after standardizing the process of detection in the laboratory. Each sample location should have the latitude and longitude, so that these may be plotted in the georeferenced map with help of GIS software.

4.3.8   Biogeochemical Sampling

The fundamental concepts of bio-indicators, chiefly plants and termite mounds, to locate ground water and mineral resources in arid and semi-arid regions were described by the ancient Sanskrit text "Brihat Samhita" by Varahamihira

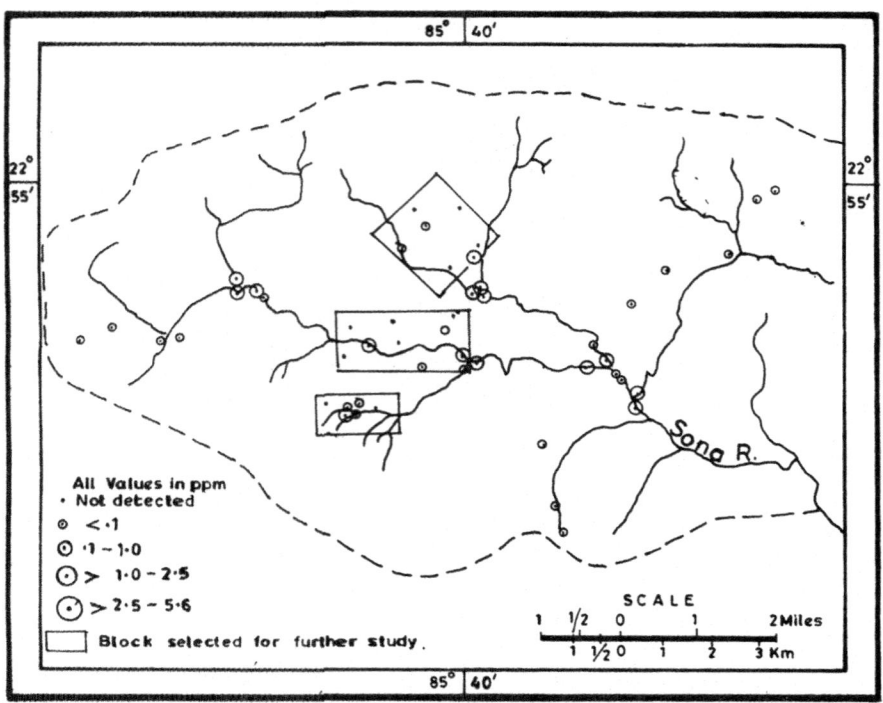

Fig. 4.6 Distribution of samples from the Sonapet Valley area with their gold contents

(A.D. 505-587). Maintenance of constantly high relative humidity in the termite mounds is an essential prerequisite for the very survival of the species in arid and semi-arid regions (Prasad, 1987). The termites penetrate deep down into the subsoil to reach the water table. Along with the soil particles for the construction of the mounds and ground water taken out by the termites, ore elements present in the country rocks around the water table are also brought to the surface. Hence the mounds show anomalous concentration of ore elements from the adjacent soils if there are any subjacent ore bodies. In areas with thick laterite overburden or where the overburden is too deep for auger probes, the application of biogeochemical prospecting adopted by sampling of leaf, bark and root of the prevalent plants or tall trees of the area is taken up (Brooks, 1972; Chaffee and Gale, 1976). These samples are dried and burnt to ashes. Ashes are sent to the chemical laboratory for analysis of different trace elements. Table 4.1 shows the trace element content in granite, associated soils and plants.

The age of the plants are also taken into account while sampling plant parts. Even seasons have effect on the dispersion of essential trace elements in the plants. Accordingly, different plant samples are taken on one particular season. A set of

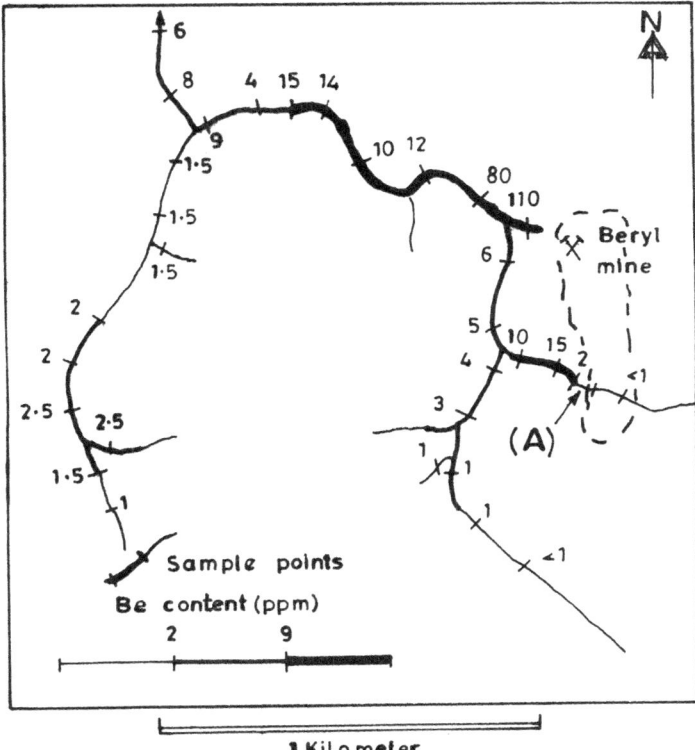

Fig. 4.7 Distribution of Be in stream sediments where (A) indicates the approximate location of Beryl bearing fertile pegmatite and the numbers by the side of sample points indicate concentration of Be ppm

data on the preferential concentration of some trace elements by different plant organs is given in Table 4.2 to give an idea in this regard. Once the data is plotted in the map, contour may be drawn to outline the anomalous area for further exploration.

4.3.9 Homogeochemical Sampling

It deals with some of the diseases in human beings caused due to excess intake/ deficiency of certain elements in the local terrain through drinking water. Thus human population are also good indicators for geochemical prospecting in certain

Table 4.1 Average contents of the elements in the 'granite' of the earth's crust, in soil and in plants (in ppm)

Element	Granite layer of earth's crust	Soil	Plants (in ash)	Conc. ratio (plant / soil)	Element	Granite layer of earth's crust	Soil	Plants (in ash)	Conc. ratio (plant/ soil)
Li	30	30	6	0.2	Cu	22	20	20	1.0
Be	2.5	3	3	1.0	Zn	51	50	900	18
B	10	10	400	40	Ga	19	20	20	1.0
F	720	200	10	0.05	As	1.6	5	0.3	0.06
Ti$^+$	3300	4600	1000	0.2	Se	0.14	0.01	?	?
V	76	80	61	0.8	Rb	180	130	120	0.9
Cr	34	50	90	1.8	Mo	1.3	3	9	3.0
Co	7.3	10	15	1.5	Ag	0.05	0.1	1	10
Ni	26	30	20	0.7	Cd	0.15	0.5	0.01	0.02
Au	0.001	0.001	0.007		Sn	2.7	4	5	1.2
Hg	0.03	0.01	0.001	0.1	I	0.5	5	50	10
Pb	16	10	10	1.0	Cs	3.8	4	4	1.0
H	2.6	1	0.5	0.5					

Table 4.2 Distribution of some trace elements in different organs of the plant

Trace element	Plant organ analysed	
	With maximum concentration of the trace element	With minimum concentration of the trace element
Pb	Roots	Bark
Zn	Leaves	Wood
Cu	Roots	Bark
Li	Leave of trees	Bark
	Parts of grasses	
Ni	Leaves	Bark

areas. Some of the diseases commonly observed and the causative element/s are stated below:

Sclerosis → Hg   Arthritis → Cd
Fluorosis → F   Lead poisoning → Pb (Plumbosis)
Goiter → I Deficiency   Silicosis → Mine Dust
Arsenosis → As   (Mostly quartz)

People suffering from such diseases give first hand information about the nature of anomalous trace element distribution within the surrounding rock, soil, water etc. of the area under investigation. Quantitative data showing the village-wise distribution of effected people may give some idea about the underlying causative mineral deposit or occurrence of the area.

All the above techniques of geochemical sampling are shown as a flow chart in Fig. 4.8.

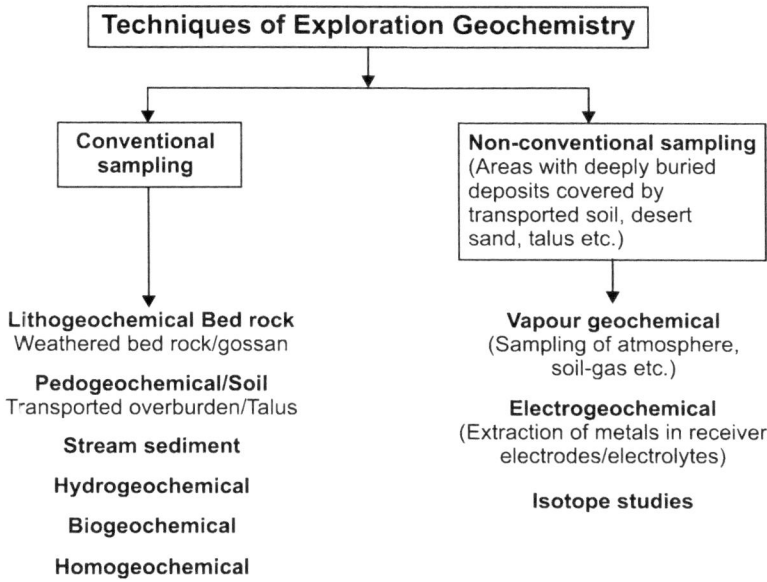

Fig. 4.8   The different techniques of exploration geochemistry

4.4   STATISTICAL ANALYSIS AND INTERPRETATION OF GEOCHEMICAL DATA

Geochemical data received from the analysis of samples collected from many parts of the world require statistical analysis for proper interpretation, which in the long run help to detect the presence of potential mineral deposit, if any. Geochemical samples, irrespective of the sample media chosen, give rise to huge volume of analytical data, normally in ppm level, with some values in ppb and rarely in percentages. This is applicable for sampling in any part of the world. Once the location co-ordinates and analytical results of these samples are stored in a database along with other geological parameters in a computer-compatible format, different types of statistical analysis may be conducted as per the requirement. But, before undertaking any EDP technique, the analytical data of geochemical samples need be processed mainly by graphical means, using statistical plots, which are discussed in subsequent paragraphs.

Scanning of the results of multi-element geochemical analyses at first sight gives some idea about the range of values of individual parameters along with their distribution pattern. But systematic statistical evaluation of such data initially requires some univariate methods of study, since thorough understanding of individual variables/parameters is essential for the interpretation of multivariate studies, especially in the case of geochemical survey and exploration. Therefore, a number of routine univariate analyses that deal with statistics of random variables are usually undertaken (Govett, 1983).

A few widely used terms may be defined before describing the commonly used statistical plots of geochemical data. The most commonly used measure of central tendency of a set of continuous data is the arithmetic mean or average, which is obtained by summing all items and dividing the same by the number of items (n) $\bar{X} = \left(\sum_{i=1}^{n} X_i \right) / n$. This 'mean' is an estimate of the true mean of the population that has been sampled. Other measures used are the 'median' and the 'mode'. The 'median' separates the upper 50% from the lower 50% of a data set. The mode is the value (or narrow range of values in the case of histogram) that is more abundant than other immediately adjacent values (or ranges of values). Dispersion of items generally is evaluated as the mean squared difference relative to the arithmetic mean, a measure called 'variance':

$$s^2 = \left(\sum_{i=1}^{n} \left(X_i - \bar{X} \right)^2 \right) / (n-1)$$

The standard deviation (s) is the square root of the variance and is the most commonly quoted measure of dispersion.

In exploration geochemistry, background is a term that tends to be loosely used implying the abundance of an element in a particular material or media. Conventionally this abundance is represented by the average (mean) content of an element. This approach is reasonable only for a normally distributed population. A better approximation to the most commonly occurring value is the 'geometric mean' (GM). The GM reduces the importance of a few high values in a sample group and is, therefore, numerically less than the arithmetic mean, making it a useful indicator of background for most geochemical data.

4.4.1 Statistical Plots of Geochemical Data

Among the varying types of graphical plots, a histogram has the advantages of providing visual information on the total range of values in a data set and the range of greatest abundance of values recorded for each class interval. If constructed with frequency as a percentage, histograms provide a means of comparing similar types of data from different sources and based on different numbers of items. However, care must be taken in the construction of a histogram so that it provides an honest representation of a data set. A class interval in a histogram should generally be in the range one-quarter to one-half the standard deviation, and the frequency scale is most useful as a percentage. The histogram should invariably record the mean, standard deviation, class interval and total number of items processed for the entire data set.

A stem-and-leaf plot using one variable at a time may also display a set of geochemical data. In this plot, actual values are plotted overcoming the need for class intervals for the histogram.

Probability graphs using probability paper is a useful practical tool in the analysis of geochemical data because of the common 'normal' or 'lognormal' character of such data. In such diagram one ordinate is either arithmetic or logarithmic scale as required and the other one is the probability scale that will plot a cumulative normal or lognormal distribution as a straight line. This type of graph paper is fairly sensitive to recognition of combinations of multiple populations, hence its use for plotting of geochemical data.

4.4.2   Recognition of Anomalies

Recognition of anomalous values are often made using frequency plots of geochemical data. Assessment of threshold from a cumulative frequency graph for a data set from a normally distributed background population is represented by the mean plus two standard deviations. This is the classic approach for calculating threshold by Hawkes and Webb (1962), but not for any and all types of geochemical surveys. To apply parametric statistics the data must be transformed to a normal distribution. A common practice is to convert the raw data to log 10. The frequency distribution that results from such conversion has a slightly negative skew, but it is much closer to a normal distribution than the arithmetic data. Log-transformation has the effect of diminishing the importance of high values (i.e., anomalous values) relatively to low values (i.e., background values). When adequate control data are available, the samples in the tail of a distribution can be shown to belong to a separate (anomalous) population that overlaps the main (background) population. In such cases, an alternative approach to log transformation is to attempt to separate the populations by the method of partitioning, so as to estimate the threshold of the lower (background) population separately. A graphical solution is the easiest approach, utilising the fact that the cumulative percentage frequency of a normally distributed population plots as a straight line on arithmetic probability paper. Deviations from a straight line generally with kinks, indicate a mixture of two or more population. There are various techniques for estimating geochemical threshold from probability plots (Lepeltier, 1969) and partitioning of populations (Sinclair, 1974).

Soviet geochemists use rock geochemistry data extensively for interpretation of geochemical result generally by means of statistical plots which indicates primary dispersion. These include use of simple arithmetic manipulations of multi-element data based on a general zoning sequence of trace elements. They frequently use "additive halos" (addition or subtraction of different element values standardised to respective background for each element), "multiplicative halos" (multiplication or division of element values) and ratios of supra-ore to sub-ore elements. The use of multi-element ratios has the advantage of minimising analytical variations (where most of the analyses are semi-quantitative spectrographic type) and reducing the effect of local reversals in zoning sequences. Point symbol maps as well as contour maps prepared from these data give good indications of anomalous zones.

4.4.3 Univariate Analysis

A thorough understanding of individual variable/parameter is essential for the interpretation of results of geochemical appraisal in any area. As such, univariate methods are fundamental to all statistically oriented geochemical studies. In many cases, the results of multivariate studies can be foreseen by a detailed univariate approach. In such case emphasis will be on data that are continuous or approximately so, such that data items can assume any value within an observed range. In general, samples must be evaluated critically as to their randomness or lack of bias.

Normally a histogram shows the frequency of occurrence of different contiguous narrow ranges of values. As the close interval becomes narrower, a smooth continuous curve may be passed through the tops of the classes with increasing ease. Such approximation of the disposition of values by a continuous curve is likely if data were continuous or nearly so. Such curves are referred to as density distributions and in practice they can be represented by specific mathematical functions. Generally, it is not sufficient to adopt an ideal density distribution for a data set by purely subjective decision. Several statistical tests exist that permit a comparison of real data with various models or hypothesis, like chi-square test, F- and t-test (vide Govett, 1983).

Among the univariate analyses, the analysis of variance includes a wide variety of important statistical procedures dependent on the fact that the total variability in a data set can be divided into parts that arise from different sources. Individual sources of variability can then be assessed relative to each other. In the evaluation of geochemical data, we generally confine our attention to simple fixed and random models that have wide application. In the model study, variations within groups of data are compared with variations between groups. If the between-group variations are greater than the within-group variations, it is concluded that mean values of the two groups differ. Such a comparison is generally made by an F-test. A common application of analysis of variance is to compare variations that arise from different identifiable sources. In geochemistry a case in point concerns the relative magnitudes of analytical error, sampling error and regional variations. In order to make this comparison, one must separate variations arising from the different sources and make necessary comparisons using the F-test.

4.4.4   Multivariate Analysis

Statistical methods that take account of the various relationships between the different variables are termed multivariate statistical analyses. The modern analytical laboratory provides an abundance of multi-element information for a geochemical survey, which will necessitate computer-aided interpretation for proper and efficient evaluation. The presence of mineralisation within a survey area will manifest itself in a geochemical survey as a variation in the context of a chemical element, indicating a geochemical condition that does not fall within the range of normal background condition. Under such condition, multiple regression analysis may be used to estimate the background concentration of a sample site for a given element on the basis of all other elements measured (Rose et al., 1970). If the actual value obtained from chemical analysis is significantly higher than the estimated background value for the element, the sample site is considered anomalous. The analysis of multi-element data from a whole geographical area is useful for the identification of broad features and will aid the estimation of the influence of one element on another, with the intention of eliminating false anomalies and also sharpening others in terms of their anomaly to background contrast. It will also help to focus attention on data associated with particular geological or physiographic units. Regression analysis is now becoming widely used to detecting those samples in which element concentration is attributable to processes of mineralisation in a set of geochemical data.

In order to be able to adequately describe the characteristics of the geochemical background population in an area, it is important that outliers (with very high values) should be identified and omitted from the data set prior to the analysis. This should preferably be done during the preliminary scanning and validation of data. It is often advantageous to split a large data set into more homogeneous subsets prior to further statistical treatment in order to avoid the problem of dealing with polymodal distributions for the individual elements.

In order to group different samples together on the basis of their similarity in terms of their compositions, there are a wide variety of different cluster analysis techniques (cf. Govett, 1983). It may, however, be noted that the applicability of many of these techniques to the very large data sets involved in normal geochemical survey work has limitations and hence should be used with proper caution. Scaling of the data for each variable, as well as any functional dependence between variables, is also important since the majority of techniques employ same kind of distance criterion for inter sample or inter-group similarity. It follows that sometimes reducing the multi-element data to similar ranges of values for each attribute is important when a distance-related criterion is used in cluster analysis, *so* that efforts of large numbers do not unduly outweigh those of small ones. In cluster analysis grouping precedes by first finding the most similar pair of samples. These then form the initial group, and all other samples are compared with it. If any sample is more similar to samples already grouped than to the rest, it is assigned to that group, otherwise a new group is formed of the next most similar pair of the samples. The final result is a tree-diagram, referred to as 'dendrogram', in which the end points of the branches represent the samples, and the height at which the branches join corresponds to the sample similarity level for admission to a pre-existing group (Fig. 2.2). The higher the branches join, the more similar are the samples. Different grouping methods (like single linkages, nearest neighbour, weighted average etc.) give rise to differences in the structure and pattern of the dendrogram tree.

While evaluating geochemical samples, it is often more helpful from the interpretational point of view to find out the mutual relationships between the different variables of individual samples. In the case of very large number of samples with multi-element determinations *"Pearson Correlation Coefficient"* between all element pairs will give much useful information on the structure of inter-element relationships. A standard formula for determination of Correlation coefficient (*'r'*) value is available in any textbook of statistics. While using correlation coefficient values, precaution should be taken to guard against spurious results. Testing its significance requires that the data elements are approximately normally distributed and the number of samples needed to reliably estimate the multivariate correlation matrix should be at least three and preferably at least ten times the number of elements involved.

4.4.5 Principal Components Analysis (PCA) and Factor Analysis (FA)

In exploration geochemistry these analyses are generally used to separate the element associations inherent in the structure of the correlation matrix, into a number of groups of elements, that together account for the greater part of the observed variability of the original data. The aim behind these being to represent the large number of elements in the original data by a smaller number of factors, each of which is a linear function of the element concentrations, thus giving a greater efficiency in terms of information compression over the original data. PCA and FA have much in common, the former is variance-oriented, while the latter is correlation-oriented. For geochemical data, PCA appears to be favourable in situations in which the range of variation of the elements is characteristic of the geochemical environment, whereas FA is favourable in situations in which element associations characterise the geochemical environment. Detailed account of the procedure with examples for both the types of analysis is given by Govett (1983).

Another important statistical technique is called *Discriminant Analysis,* which is aimed at devising an optimum set of rules for the classification of a sample into one of a number of pre-defined groups based on a number of measurements. Let us consider, for example, a number of representative samples from both gossans associated with mineralisation and barren laterites, and that Cr and Ni values have been determined for each sample. Now, it is intended to set up a rule for the allocation of a sample of unknown affinity to one or other of the two groups, based solely on its chromium and nickel values. This type of problem can be solved by one of a number of standard discriminant analysis techniques each defining the decision-making criteria in a slightly different way. The linear discriminant function most frequently used in geochemical work can be imagined as a line orthogonal to the separating boundary between the point plot of different groups. The boundary will ideally be placed so as to minimise the error rate (i.e., samples truly belonging to one group being wrongly assigned to another). Linear discriminant functions have been successfully used in exploration geochemistry by a number of workers.

Trend Surface Analysis is used in geology to separate regional trend of selected parameter from local variation. However, the terms, 'regional' and 'local' are highly subjective. The trend surface maps based on geochemical samples with known coordinates depend upon various factors such as size of the anomaly, distance between samples, geometry and size of sample grid etc. In trend surface analysis, a series of surfaces are fitted to data points. A typical programme consists of three basic parts: (i) a routine to generate the matrix of sums of power of cross products, (ii) a simultaneous equation solver of matrix inverter and (iii) a plot algorithm. The equation chosen for trend surface is so that the squared deviation from the trend is minimized. The first and second degree correlations are generally used as measure of goodness of fit. Even a random set of data generates a correlation of 0.3 for fourth degree polynomials. A value of 0.7 is generally considered to be good fit. A statistical test of goodness of fit is done by comparing the variance due to regression to the variance due to deviation from the trend with an ANOVA programme routine.

For structural and basinal studies three dimensional trend surface analyses are carried out comprising X and Y co-ordinates and value of elevation or depth to certain datum plane. However, in geochemical studies, four dimensional trend analyses are common. This consists of X, Y and Z co-ordinates and any other variable of interest, such as chemical analyses of a particular component/element. The completed analysis has the form of a solid containing nested contour sheets or envelopes. A traditional method of construction is to create a series of maps, at different levels, i.e., a series of horizontal cross sections through the trend envelopes and then to assemble the whole by stacking the maps. The following precautions should be observed while undertaking the analysis:

1. Number of control points should be sufficiently large so that meaningful statistical tests can be run.
2. To overcome edge effect, data should be collected over an area greater than the size of the area under analysis.
3. The distribution of sample points should be random.

In conclusion, it may be said that the above mentioned techniques are sufficient to (i) give a good understanding of the data, (ii) help choose a suitable method of statistical analysis of the data and (iii) highlight anomalous observations whose effects on latter analysis may be useful.

Chapter 5
NON-CONVENTIONAL TECHNIQUES FOR CONCEALED DEPOSITS

5.1   INTRODUCTION

Non-conventional techniques of exploration specially for concealed mineral deposits, present in any part of the world, require some new approaches that may be applied for this purpose. In this regard, Geological Survey of India through its Field Technique Research Unit (FTRU) tried to develop some new techniques of exploration of concealed mineral deposits since 70s of last century. GSI tried to standardize some such new techniques which will be discussed in this chapter. Most of the mineral deposits of the world are generally not exposed at the earth's surface. These may be partly or wholly concealed by thick residual soil or buried beneath the rock formation. To locate these deposits, some conventional geochemical exploration techniques are generally employed to locate these deposits based on systematic measurement of one or more chemical property in different sample media like rock, soil, weathered profile, glacial debris, stream sediment, water, plant etc. as described earlier.

The chemical property most commonly measured is the content of the element of interest ("Key element" or "Pathfinder elements") that are likely to be associated with the mineral deposit. The analytical results may indicate the geochemical anomaly, if any, in the area that may ultimately guide to locate the mineral deposit. In areas where the conventional geochemical exploration techniques are not applicable (such as, areas with deeply buried deposits covered by transported soil, alluvium, desert sand, talus material etc.), some non-conventional techniques are adopted which are described here that may be applicable for any part of the world. These techniques are based on methodology of detection of concealed base metal sulphide and related deposits applying the concepts of vapour geochemistry,

© Capital Publishing Company, New Delhi, India 2020   117
A. K. Talapatra, *Geochemical Exploration and Modelling of Concealed Mineral Deposits*, https://doi.org/10.1007/978-3-030-48756-0_5

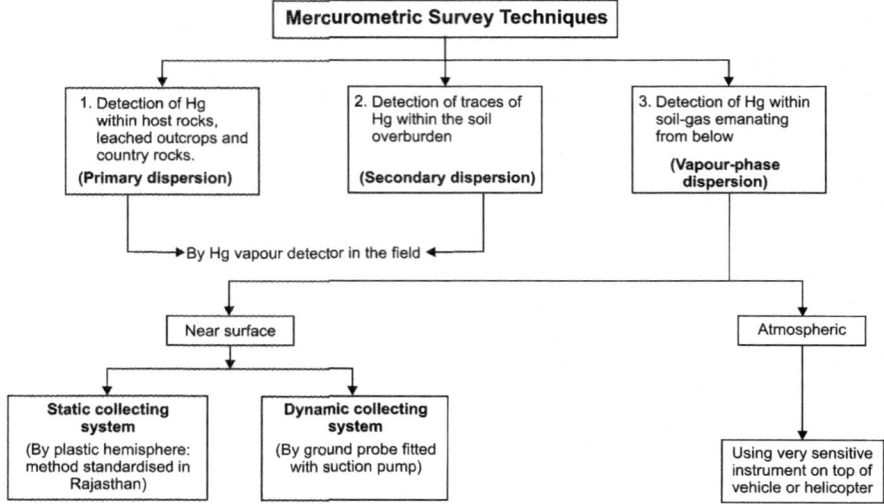

Fig. 5.1 Various types of mercurometric survey techniques

electro-geochemistry and isotope geology (Talapatra, 1994). These non-conventional techniques are capable of detecting the concealed lode zone in extension areas of known deposits as well as in virgin areas, which are covered by transported soil, desert sand, etc. Figure 4.8 in Chap. 4 shows the different techniques of exploration geochemistry including the non-conventional types.

5.2   APPLICATION OF VAPOUR GEOCHEMISTRY

Research and development unit of GSI mentioned earlier tried to develop some methodology and technique for discovering concealed mineral deposits in land areas with the help of vapour phase geochemistry detecting different types of soil gas like Hg, SO_2 etc. for discovering concealed base metal sulphide deposits. In addition, exploration techniques utilizing principles of electrochemistry has also been developed successfully by GSI for searching base-metal sulphides in parts of some mineral belts of India. The techniques utilize the electro-positive property of the metallic ions, where a large section of the ground is artificially electrolised, which will be discussed subsequently. Locating new targets of ore deposits from the covered areas beneath barren, exotic or transported overburden is one of the most difficult tasks for present day mineral exploration. The deposits in the close vicinity of outcropping gossans, old workings, slag dumps, secondary surface encrustations and stains, have already been fully explored in most of the countries. As such discovery of new mineral deposit with the help of conventional methods described above is quite difficult. Non-conventional geochemical techniques applying soil-gas geochemistry, electrogeochemistry etc. under such circumstances can be used for exploring base metal sulphides by trapping soil gas. Even in ancient times, the association of peculiar odours with certain mineral deposits has been noted and in Scandinavia, sometime dogs have been used in locating mineralized boulders in glacial till. It must, however, be mentioned that vapour surveys suffer from several limitations, such as, (a) the results of vapour surveys are difficult to repeat, (b) vapours present in the soil may not be detectable in the air and (c) gases collected in the field may be from sources other than underlying bed rocks. Normally, the gases released in the air may be from oxidation of ore deposits lying below, through pore spaces of rocks, joint, fracture, fault etc. where degassing activity starts and continuously flow through the overburden. Table 5.1 shows a list of gases and associated mineral deposits commonly encountered in the field.

The most widely used vapour surveys have so far been mercurometric and sulphur dioxide surveys (Rouse and Stevens, 1971; McNerney and Buseek, 1973).

5.2.1   Mercurometric Survey Techniques

Exploration of buried base-metal sulphide mineralization by conventional geochemical methods sometimes becomes difficult in certain field conditions where the host rocks of ore minerals and associated country rocks are covered by transported soil, sand (alluvial/desertic), talus or similar other materials. In such a condition mercurometric survey techniques for exploration of concealed mineralisation work as a fairly satisfactory tool (McCarthy Jr., 1972). The high volatility of mercury is responsible for its wide dispersion in both geological and biological environments. Significant presence of mercury in the rocks and surficial materials of some base

Table 5.1 Vapour indicators of ore deposits

Vapour	Type of deposits
Mercury	Ag-Pb-Zn sulphides; Zn-Cu sulphides; U-Au, Sn-Mo and polymetallic Hg, As, Sb, Bi, Cu ores; pyrites
	All sulphide deposits
Sulphur dioxide	All sulphide deposits
Hydrogen sulphides	All sulphide ores; Au ores
Carbon dioxide and oxygen	Pb-Zn sulphides; porphyry copper deposits
Halogens and halides	U-Ra ores; Hg sulphides; potash deposits
Noble gases	Possibly all sulphides (Pb, Cu, Ag, Ni, Co etc.), Au-As, HgAsH
Organo-metallic, such as $(CH_2)_2$	Nitrate deposits
Nitrogen oxides	

metal sulphide and mercury provinces, and its tendency to accumulate in the tissues of certain living organisms in traces, makes it an element of considerable geochemical interest. But, due to non-availability of reliable and quick method of analysis for this element at trace level in geological samples, mercurometric survey could hardly be widely adopted in geochemical exploration in the past. Subsequent developments of analytical methods for quick analysis of mercury (Ure and Shand, 1974; Ure, 1975; Chilov, 1975) viz. cold-vapour atomic absorption method etc. have opened up a new vista for exploration of mineral deposits, associated with mercury dispersion.

It has been observed that sulphides of noble metals and chalcophile elements like Zn, Pb, Cu etc. are associated with considerable amount of mercury (Saukov, 1946; Shipulin et al., 1973). Highest correlation of Hg has been noted with Zn in Archaean samples (Cameron and Jonasson, 1972), while sympathetic relations of Hg with other elements like Pb, Ag, Cu etc. are also noted in a number of mineralized belts. A fairly detailed account on the geochemistry of mercury has been given by Jonasson and Boyle (1972) who have outlined a possible mechanism of mercury released from sulphides in the zone of oxidation. According to them, Hg ions are liberated from sulphides by the reaction of ground or surface waters containing natural oxidants such as ferric ion, by bacterial attack, or by even more exotic oxidants such as chlorine, which can be derived from the interaction of manganese dioxide and acidic waters. McNerney and Buseek (1973) also showed that native mercury can be generated by natural processes from any sulphide containing minor amount of mercury, where oxygen is available.

The Hg-vapour thus released can migrate through a considerable thickness of the country rocks and overburden along the weak planes, pore spaces etc. During secondary dispersion in vapour phase, mercury partly gets fixed within the silt and clay fraction of the soil and the remaining portion is carried as soil-gas which may produce pronounced local anomalies of Hg directly above ore deposits (McNerney and Buseek, 1973). Mercury aureoles are also known to occur for considerable distances as far as two kilometres away from the primary source (Ozerova, 1962). Thus, from the available data, it may be suggested that the various types of mercury detection technique may be applied for delimiting specific ore bodies as well as general target areas (Fig. 5.1).

5.2.1.1   Methodology

Mercurometric survey techniques for exploration of concealed deposits thus, include methods of detecting traces of mercury present within the soil overburden, bed rocks, leached outcrops etc. as well as within the soil-gas emanating from below (including both near-surface and atmospheric Hg-vapour). On the spot detection of Hg in soil samples was carried out in parts of Rajasthan by a portable mercury-vapour detector, the details of which will be discussed here. In the other technique of soil-gas sampling, mercury is made to form amalgam with noble metal foil or wire of Au, Ag or Pt traps. Collection of Hg from the soil-gas phase can be effectively done by means of two devices: (1) by a static collecting system in which a colourless and transparent plastic hemisphere placed over soil at the site of sampling is used to funnel air with soil gas through the noble metal collectors made up of thin wire guage and (2) by a dynamic collecting system in which a special type of ground probe is utilized to suck and trap such soil gas through noble-metal collectors with the help of a pump (McNerney and Buseek, 1973).

Talapatra and Bose (1979) successfully adopted a static collecting system with slight modifications using indigenous materials and equipments (Anon, 1976) for trapping Hg-vapour from soil-gas (Fig. 5.2), the details of which are discussed later. The methods of mercurometric survey adopted by the authors were initially stan-dardised in known mineralized belts of Dariba-Rajpura and Khetri areas of Rajasthan. A short geological account of these two areas is given below.

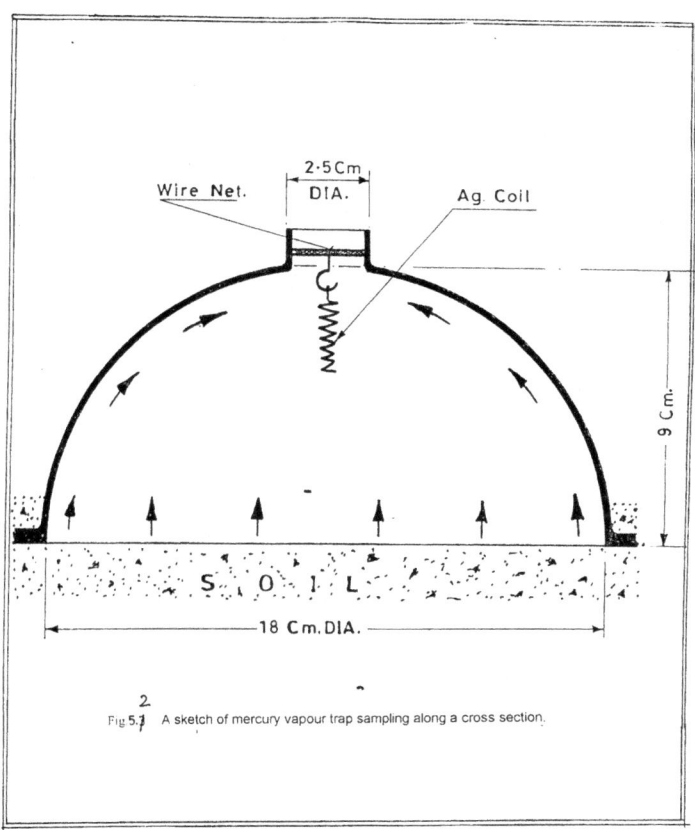

Fig 5.2   A sketch of mercury vapour trap sampling along a cross section.

Fig. 5.2 A sketch of mercury vapour trap sampling along a cross section

5.2.2 Geological Setting and Nature of Mineralisation of Study Areas

5.2.2.1 Dariba-Rajpura Mineralized Belt

The belt of polymetallic sulphides are associated with high grade metamorphic rocks of Pre-Aravalli Super Group of Precambrian age, that form a part of the western limb of the north-easterly plunging macro-syncline (Raja Rao et al., 1970), whose closure lies to the south of the area near Bhinder (24°30′ : 74°12′). The rock types of the area are garnet-staurolite-kyanite-biotite schists, graphitic mica-schist and calc-silicate rocks with bands of dolomite, orthoquartzite and amphibolitic group. *In situ* hard silicified gossan with spongy boxworks and various shades of colour (cf. Talapatra, 1979) is very well developed along graphite schist-dolomite contact zone between Dariba and Rajpura villages as a thick continuous band (Fig. 5.3). Another small band of typical gossan about 600 m long is present within graphitic mica-schist, east of Dariba. Extensive old workings, slag and mine dumps heaps are found all along these gossan zones. Two distinct polymetallic sulphide lodes have been located below these two gossan zones, occurring as steeply dipping concordant bands/lenses. Primary sulphide minerals present in these lode zones include sphalerite, galena, chalcopyrite, pyrite, pyrrhotite etc. with some sulphosalts like geochronite, boulangite, tetrahedrite, tennantite etc. (cf. Poddar, 1972, 1974).

It was initially believed on the basis of available surface evidences that mineralisation in this belt is of epigenetic hydrothermal type and is localized along structural dislocations like fault zone in high grade metamorphic rocks (Poddar and Mathur, 1963; Poddar and Chatterjee, 1966; Raja Rao et al., 1970; Raja Rao and Chatterjee, 1972). Subsequent work including subsurface exploration by numerous drill holes and pilot mining demonstrated that the sulphide assemblages are restricted to definite lithological units and are characterized by excellent bedded geometry having some distinct imprints of metamorphic recrystallisation, mobilization and deformation. These have led Poddar (1974) to postulate a sedimentary metamorphic hypothesis for these polymetallic sulphides. Recent workers have studied in detail and compared all the sediment hosted Pb-Zn deposits of Rajasthan located within North-western Indian Shield and grouped Rajpura Dariba deposit as metamorphosed SEDEX type. The deposit has assumed special significance because it hosts probably one of the oldest (1800 Ma) SEDEX type Pb-Zn deposits in the world (Deb and Pal, 2004).

5.2.2.2 Khetri-Banwas Mineralized Belt

This belt of base-metal sulphides occurring close to Khetri and Banwas extends within metamorphosed rocks of Delhi Super Group of Precambrian age comprising the older Alwar Series which is dominantly arenaceous represented by a thick pile

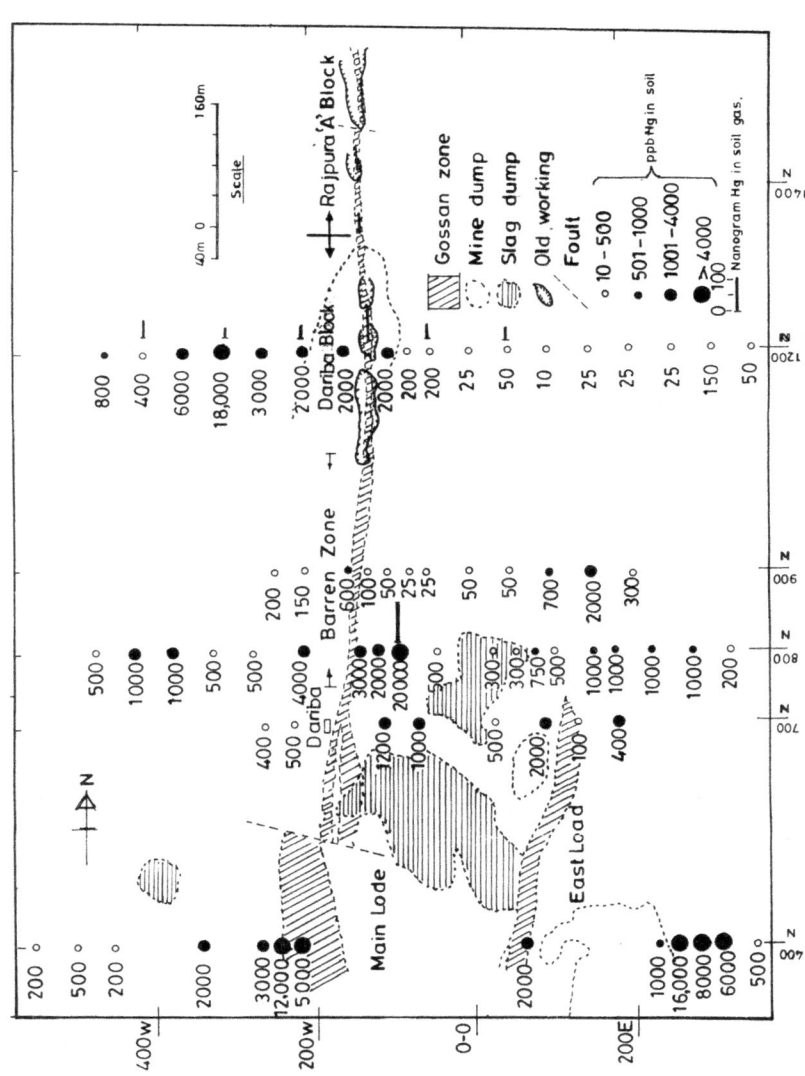

Fig. 5.3 Dispersion of mercury in soil and soil-gas with respect to Zn-Pb mineralization in parts of Dariba and Rajpura "A" block, Udaipur district, Rajasthan, India. (After Poddar and Chatterjee, 1966)

of metasediments varying in composition from orthoquartzite to arkose locally interbedded with phyllites, carbon phyllites, schists, marbels and amphibole quartzites. The Ajabgar Series is composed of metamorphosed argillites consisting of various types of schists and phyllites with local bands of marble and calc-silicate. The contact between the two series is gradational (Roy Chowdhury et al., 1968). The metasediments of this belt have undergone repeated fold movements on a large scale due to which a series of doubly plunging folds are conspicuously present (Das Gupta, 1968). The mineralisation appears to be confined within the border zones of Ajabgarh and Alwar series. The sulphide lode occurs in this belt as a number of detached lensoid bodies of different scales with intervening lean/barren portions both along the strike and dip direction of the lode. The ore body consists, mainly, of pyrite, pyrrhotite and chalcopyrite with other minor sulphides namely sphalerite, galena, arsenopyrite etc. Das Gupta (1964) described this as an epigenetic sulphide deposit occurring as fissure filling and replacement along shear zones within rocks of the amphibolite facies, and that the ore forming fluids were derived from the hydrothermal solution mostly, at hypothermal to mesothermal conditions. Later, Das Gupta (1974) slightly modified his earlier ideas and listed the features supporting the epigenetic origin of these sulphides along with some syngenetic characters, preserved in this belt, which together suggest a dual mode of origin, though, according to him, the available data are still in favour of an epigenetic type.

5.2.3   Field Detection of Mercury in Soil

Systematic mercurometric survey of soil samples was taken up along Dariba-Rajpura mineralized belt, since abnormallous high values of Hg (by spectroscopic analysis) had been reported from sphalerite and associated sulphides of this belt (Raja Rao et al., 1970). Sampling of soil for field detection of Hg was carried out along a number of E-W trending lines across the gossan zones of Dariba Block (Fig. 5.3) and one such line along the central half of each of the three Blocks of Rajpura area. Soil samples were collected preferably from 'B' horizon by making small pits about 20-30 cm deep from the surface. As the area shows evidences of ancient mining activity, care was taken to avoid the old workings, slag dumps, and mine as far as practicable. From the air dried samples hard concretions were removed and then sieved to –40+80 mesh for geochemical analysis of Hg in the field and spectrographic analysis of other elements at the chemical laboratory of G.S.I., Central Head Quarters.

5.2.3.1   Rapid Field Analytical Procedure

In the field, mercury in the soil samples was geochemically analysed with the help of a mercury-vapour detector (Lemaire Instrument Co., Canada, model No. 300), which is fitted with a low pressure mercury lamp and an absorption cell. A scoped

sample in the range 0.01 gm to 1.0 gm is mixed with equal amount of barium peroxide and heated in a metal bulb connected to a hand operated suction-pump. The
vapourised mercury liberated from the heated sample is slowly sucked in by the
pump, and filtered through a special type of filter paper for removing dust particles
and moisture. The purified vapour collected in the pump is then flowed through the
absorption cell where absorption of the resonance line of mercury causes a drop in
the detection signal which in turn is recorded as a scale reading. This vapour detector is small and battery operated. Its detection range is between 5 and 20,000 ppb.
Barium peroxide and iron powders were mixed with the sample to minimize the
error. Even then, duplicate readings were usually taken to minimize the error arising
out of other spectral interferences. Results of Hg values adsorped within soil fraction of Darba-Rajpura 'A' Block are given below.

5.2.3.2 Results of Mercurometric Survey of Soil

Mercury values in the soil samples of this area varied from 10 ppb to 20,000 ppb;
the higher values are generally recorded from near the lode zones, excepting some
anomalous values close to the mine dump areas (cf. Fig. 5.3). Along 800 N line,
very high mercury values in soil (about 20,000 ppb) have been recorded adjacent to
the east of the 'Barren Zone' of the Main Lode. Mercury-vapour trap sample collected from this spot also shows significantly high value (125 ng), which may be due
to either the presence of Hg and associated elements in this part at depth or to some
local structural causes. Further east, high values of Hg (around 1000 ppb) in soil are
spread over 120 m portion of the same line. High values of Hg with anomalously
high values of Pb (0.36% to 2.05%) in soil samples along 900 N line have been
noted along the possible extension zones of both the lodes. The same along 1200 N
line does not show any significant value along the probable extension areas of the
East Lode, while in the western part of this line, anomalous values have been
recorded close to the Main Lode similar to those of other lines. But, some high values of Hg (6000-18,000 ppb) have also been noted in this line around 300 W, which
may indicate a local enrichment gentle westerly sloping ground due to contamination of mine dump materials. This may as well reflect the presence of a separate
mineralized lens or pocket of sulphide rich in Hg and associated elements. But,
mercury-vapour trap analysis from this part however does not corroborate the latter
suggestion. In Rajpura area, Hg values in soil are found to range from 25-300 ppb,
higher values always being close to the known mineralized zone represented by gossan outcrop excepting some abruptly high values (1000-1600 ppb) near the mine
dump area in Rajpura 'B' Block.

5.2.3.3   Applicability of the Technique

Mercurometric soil survey technique, thus appears to be useful for exploration of base metal deposits by delineating the soil-mercury haloes in virgin as well as in the soil covered extensions of known mineralisation. In this method detection of Hg fixed in soil can be carried out in the field during sampling, and thereby subsequent programme of sampling can be modified to delineate the target area. Moreover, the various storage problems of mercury-bearing samples can be avoided by this method. After the detection of Hg with the help of vapour detector the soil samples may be further analysed for base metals viz. Cu, Pb, Zn, Ni, Co etc. at the laboratory to bring out the interrelationships of the elements. This method will not possibly be suitable where mineralization is covered with alluvium, transported sandy soil or airborne desertic sand unless degassing of mercury vapour flows steadily. In areas where continuous rock outcrops are exposed, representative samples of rock ground to –40 mesh may be analysed in the same way by portable mercury detector for demarcating the mercury haloes.

5.2.4   Technique of Mercury-vapour Sampling from Soil-gas

In this method, as mentioned earlier, a clear plastic hemisphere approximately 18 cm in diameter with a 2.5 cm opening at the top (Talapatra and Bose, 1979) was placed over the site of sampling (Fig. 5.2) to funnel soil gas through a 30 cm long pure silver wire (1 mm diameter) spiral, with peripheral diameter of about 1.0 cm and length of 3.5 cm from the tip of the inverted 'V' shaped hook. Silver has been used as a trapping material for being the cheapest among the noble metals that forms amalgam with mercury, which is easy to procure and convenient for handling in field conditions. The hemisphere with the Hg-vapour trap was generally placed on a level ground after scraping out the top 1 cm of soil, which was lightly piled and pressed around the base of the hemisphere (Fig. 5.2). The top of the hemisphere was covered with a porous paper and a rubber band. After a fixed period of time ranging from 1 to 2 hours (depending upon the area and nature of mineralization), the silver wire was removed cautiously by means of a fork and immediately sealed in a numbered airtight thick polythene envelope. In order to standardise the method of sampling, various trial runs changing the period of exposure of the weir as well as different time periods of the day was carried out. It was observed that samples taken between 8 AM and 2 PM are absorbing good amount of Hg. The silver weir exposed for two hours or more shows good results. About 10 to 15 plastic hemispheres were placed simultaneously along the traverse lines across the mineralized zone at a suitable interval (Fig. 5.3) and the samples were collected preferably between 8 A.M. and 11 A.M. every day in order to minimize the effect of atmospheric variations of temperature and pressure due to diurnal change. The trap samples were then sent to the laboratory for quick analysis. The results of various trial runs taken for standardising the method of sampling will be discussed later.

5.2.4.1 Analytical Procedure

In the laboratory, mercury in the silver traps were determined by Atomic Absorption Spectrometry after vaporization of mercury by pyrolysis using two different methods, namely, one for determining trapped mercury in the range of 3 to 125 nanogram (Direct method) and the other for the values ranging from 25 to 500 nanogram (Method through absorption in acidic permanganate solution), as detailed below.

Direct Method: In this method, the silver trap sample collected from the field was placed in a silica tube. This tube was previously positioned inside a small size movable tube furnace. The position of the silica tube was kept fixed. The trap (in this case spiral Ag wire) was heated at 500 °C by placing the preheated furnace over the position of the trap. The vaporized mercury was then directly taken to a quartz absorption cell placed along the light path of mercury line with the help of constant flow of air (2 lit/min.), which was ultimately vented out of the absorption cell. The absorption was displayed by a strip chart recorder fitted in Perkin Elmer Atomic Absorption 303 model. The absorption line selected was 2537 A°.

Indirect Method (through absorption in permanganate solution): This method requires two stages of experimental work. In the first stage the mercury in the silver trap was vaporised as in the direct method. The vaporised mercury was simultaneously carried by constant flow of air (900 ml/min) for getting absorbed in 40 ml of freshly prepared acidic potassium permanganate solution kept in a sintered glass bubbler (Quickfit, porosity No. 2). This solution was prepared by adding 50 ml of a mixture of conc. nitric acid (redistilled) and conc. sulphuric acid in 1:1 ratio to 400 ml of demineralised water, cooled, and then adding 50 ml 6% potassium permanganate solution. In the second stage of the experimental work, this permanganate solution was treated with 20 ml of a reducing solution prepared as follows: 10 ml of a mixture of conc. nitric acid and conc. sulphuric acid in 1:1 was added to 140 ml demineralised water and then after cooling, 50 ml 10% NaCl and 20 ml 25% hydroxyl ammonium sulphate were mixed with it. Mercury in this solution was determined by cold vapour AA' technique after partitioning mercury vapour in the liquid and air phase (Ure and Shand, 1974). The trap samples collected from Khetri belt during December, 1975 and those collected from Dariba-Rajpura belt were analysed by this method.

5.2.4.2 Results of Mercury-vapour Trap Analysis

The results of analysis of the samples collected show that Hg values in the traps range from 3 to 245 ng, the anomalously high values mostly remarkably coinciding with the position of the underlying lode zones in the respective areas, the details of which are given below.

Dariba-Rajpura area: Six trap samples were collected along 1200 N line and four others from the gossan zone along the Main Lode running N-S. All these silver traps were exposed for two hours between 8 and 11 A.M. It is evident from the analytical results that the samples collected from near the Main Lode zone (vide

Fig. 5.3) show distinctly anomalous values. A few sets of trap samples were also collected during the different hours of the day from the same spot giving identical exposure times as well as varying the time of exposures in different days. The results indicate that the amount of Hg-vapour trapped in the sample increases with time of exposure around mid-day but the increase is not always strictly proportionate. For example, a set of samples collected from the small old working situated in the southern part of Rajpura 'A' Block showed that the traps exposed between 7 & 8 A.M, 8 & 10 A.M. and 10 A.M. & 1 P.M. gave Hg values of 35, 70 and 205 ng, respectively. Another set of samples collected from the same spot two days later exposing the traps for one hour only at 7, 8 and 10 A.M. gave values like 25, 30 and 70 ng respectively. This further confirms the observations made earlier that the maximum degassing of Hg vapour takes place near the mid-day as a result of pressure change with rise of temperature.

Madan-Kudan and Banwas area: In this belt Hg-vapour trap samples were collected both from the known mineralized portions as well as from the areas where concealed mineralization occurs along its extension areas below the sandy soil, wind-borne sand and alluvial cover (Fig. 5.4). One trap sample was collected right

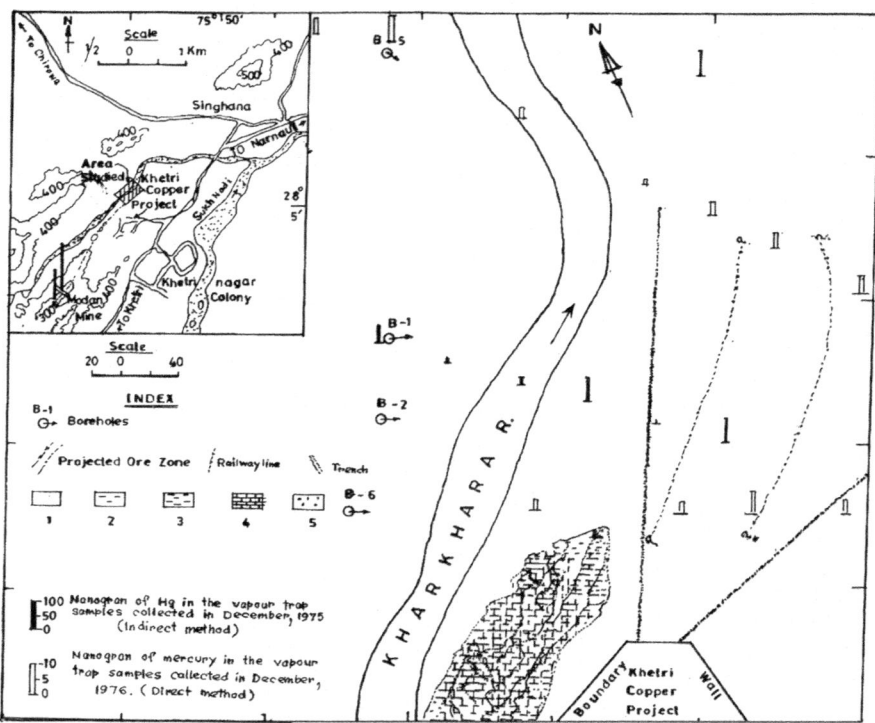

Fig. 5.4 The results of Hg-vapour analysis from Banwas Block, Khetri Cu Belt, Rajasthan with an inset location map. 1. Sandy soil, sand and alluvium, 2. Amphibolite, 3. Andalusite-Biotite Schist, 4. Silicious Dolomite Marble, and 5. Amphibolitic Quartzite/Quartzite

from the soil covered old working of Madan mine which recorded 120 ng of Hg. Four other trap samples were collected from the NW-SE trending trench within a horizontal distance of 100 m. Mercury values in these samples varied from 105 to 245 ng. Highest value being recorded by a sample collected from the central part of the trench. Alignment of the trench, with average mercury value of the four samples have been shown in the inset map of Fig. 5.4. Location of two sets of samples collected from the intervening block at Banwas sector during two successive years have also been shown in this map with mercury values. High values of Hg fit well with the projected ore zone revealed by subsurface exploration work (Bore hole points B-1 etc. in Fig. 5.4) carried out subsequently by A.M.S.E. Wing, GSI (Sarma, pers. comm.). However, presence of some high values of trap samples on the hanging wall side of the lode to the north-west, far away from the ore zone suggests that close spaced vapour trap sampling is necessary in this area for delineating the other lode zones, if any. The analytical results indicated that the amount of mercury trapped in samples collected in different periods of the year from the same locality may be quite different. Hence relative contrast of the Hg values i.e anomalous values, in a set of samples collected at a particular period, appears to be more important than their absolute values. Hence detecting the anomalous zone is important in mercury vapour trap sampling.

5.2.4.3   Applicability and Limitations of the Technique

Mercury-vapour trap sampling technique described above, has been proved to be very useful for exploration of concealed deposits, specially in arid to semi-arid areas covered by transported sandy soil, alluvial/desertic sand, talus etc., where conventional methods of geochemical exploration are not applicable. The technique will be especially suitable in desertic areas where extension areas of known mineralized belts or areas having characteristic regional geological setting suggestive of the presence of sulphide mineralization, are covered by wind-borne sand. McCarthy et al. (1969) successfully located the disseminated gold deposit in bedrock at the Cortez mine area, Nevada through 30 m of gravel overburden, based on the anomalous values of Hg in vapour trap samples. Even, ore deposits buried by more than hundred metres of post-mineral overburden have been detected here. In the present area of investigation, the depth of sand cover was not always known every where, but some of the drill holes in Banwas-Singhana area penetrated about 20-30 m of sand and loose formation (Basu, pers. comm.) where vapour trap samples from the ground surface gave significant results. This indicates that Hg-vapour has migrated through considerable thickness of transported post-mineral overburden. Moreover, it has been proved beyond dispute that once the Hg-vapour gradient to the surface is established, the thickness of over-burden or its character is not of much importance, provided it is porous and the degassing of mercury is continuous in that section (cf. McNerney and Buseek, 1973).

Some of the trial experiments showed that the silver wires should not preferably be reused, since it is practically not possible to drive out completely the amalgamated mercury from the wire, even if it is heated repeatedly. However fresh wires made out of remelting of the silver wires can be used for collecting Hg vapour trap samples. These experiments also revealed that a number of blank 'silver wire' for each set of samples should invariably be analysed for correction from blank values, if any. During the present investigation, most of the Hg-vapour trap samples were collected only in winter seasons (between December and February). Hence, more R&D work taking samples with different time periods round the year is neces-sary to show the variation of Hg-vapour degassing process in the different periods of the year.

Simplicity of this technique and its low cost of sampling are the added advantage over other vapour-phase geochemical methods, provided proper laboratory facilities are available for quick and precise analysis of Hg in the trap samples. Out of the two analytical methods, the 'Direct method' is suitable for trap samples which are pre-sumed to have lower concentration of mercury, such as those collected from virgin areas with concealed mineralization undisturbed by ancient mining activity. The other method will be useful in the extension areas of known mineralized belts with local mining activities or areas with known high mercury values, where the trap samples are likely to show high concentration of mercury. Moreover, Hg-vapour anomalies obtained from such silver sampling could be correlated to the underlying sulphide source quite easily, compared to the anomalies in soil and rock samples due to contamination with mine dumps etc., specially, the values of other pathfinder elements, as observed in both the areas studied. Such anomalies may sometimes be helpful in areas having sulphide deposits, to trace out the concealed fault or fracture zones and to establish the age relationship of these with the sulphide mineralization as has been done in parts of Yugoslavia (cf, Shipulin et al., 1973). In this connection it may be noted that soil-gas sampling gives stronger gas-anomaly for Hg and other gases than that given by air samples at greater altitudes. Thus, considering all these, the technique of Hg-vapour trap sampling discussed here rightly deserves wide application for exploration of concealed sulphide deposits in areas with thick over-burden present in different parts of India. This will, however, require proper experi-mentations by detailed sampling for standardization of the analytical technique in respective areas.

5.2.4.4   Conclusion

Anomalous concentrations of mercury in soil and soil-gas along some of the miner-alized belts of Rajasthan have been noted by mercurometric survey techniques. These anomalies correlate well with the underlying sulphide lode zones of these areas, some of which are deeply buried under alluvial/desertic sand. Thus, the tech-niques detailed above may be applied as very useful tools for exploration of con-cealed base-metal sulphide mineralization; the Hg-vapour trap sampling technique in particular appears to be most suitable in desertic areas with thick post-mineral overburden. Further detailed work in other parts of India is expected to give more information on the applicability of these techniques.

5.2.5   Sulphur Dioxide Soil-gas Sampling Technique

The Field Technique Research Unit of GSI tried to develop indigenous instruments for collecting SO_2 soil-gas samples from base-metal sulphide fields in late 70s of last century. They were successful in detecting the soil gas samples released from the underlying base-metal sulphides with the establishment of a steady gradient in the field. This will be described in the following paragraphs. The oxidation of sulphide deposits appears to generate SO_2 as a dominant gas phase along with Hg and other gases, mentioned above. With adequate permeability of the overburden, these gases will migrate to an area of lower pressure and concentration, which is generally towards the surface. Thus, the SO_2 and other associated gases when released with the establishment of a steady gradient, the soil-gas samples collected from above the mineralized zones may reflect sulphide mineralisation even lying at great depth. The application of sulphur dioxide gas surveys in prospecting was first recognized in USSR (1961). A comprehensive work evaluating the success of exploration methods using sulphur dioxide geochemistry giving some case studies on base-metal sulphide deposits of Canada and USA was subsequently presented by Rouse and Stevens (1971). A preliminary evaluation of the use of SO_2 as a prospecting tool for sulphide mineralisation in different parts of Australia has been outlined by Davy and Stokes (1976). Vapour surveys conducted by these workers have shown the presence of anomalous values in the order of 25 to 40 ppb SO_2 over deposits containing pyrite and disseminated sulphides, as against 1 to 10 ppb background values. However, it has been observed that the nature and magnitude of SO_2 anomalies around sulphide deposits are a function of a number of parameters like geology, target size, soil moisture and climatic conditions along with the factors that control the generation, migration and localization of soil gases. Industrial SO_2 pollution can also mask the anomalies partly.

5.2.5.1   Sampling of SO_2 from Soil-gas

In the SO_2 soil-gas sampling method, the soil-gas is sucked by a low capacity suction pump from small holes or pits and passed through sodium tetrachloromercurate solution. The SO_2 gas absorbed in this solution forms a stable dischlorosulphitomercurate. The amount of trapped SO_2 is then estimated by the red-violet colour produced by addition of pararosaniline hydrochloride with the help of an absorption meter or a spectrophotometer.

In order to standardize the method of soil-gas collection, SO_2 was artificially generated under laboratory condition by adding about 1.0 cc. of 1.1N HCl in 1 gm of potassium meta-bisulphite solution. The SO_2 gas thus generated was absorbed in 25 cc. of freshly prepared 0.1N sodium tetrachloromercurate solution kept in glass bubblers. By varying the strength of potassium metabisulphite solution and the period of suction, different sets of artificially generated SO_2 gas samples were collected and analysis were done at the Chemical Laboratory, AMSE Wing, Bangalore for analysis (vide Talapatra et al., 1981).

In the field, SO_2 in soil-gas samples were collected from small pits, about a metre deep by inserting an inverted plastic funnel of 10 cm diameter placed slightly above the base of the pit. The funnel was fitted with a polythene pipe of 1 cm diameter, which passes through a plastic sheet (2 mm thick) that seals the pit from atmospheric air as shown in Fig. 5.5. The pipe is joined with one end of the ancillary attachment for filtering dust particles and absorbing soil moisture. The attachment has provision to hold silica gel in a small chamber at one end and filter paper at the other. This end of the attachment is connected with a low capacity 6 volt suction pump through a glass impinger (bubbler). In order to regulate the flow of gas and measure the volume of air sucked in during sampling, a rotameter may be placed in between the glass impinger and pump. The pump is operated in the field by a 12 volt battery (Fig. 5.5). The SO_2 present in the soil-gas is absorbed in 25 ml of sodium tetrachloromercurate solution kept in glass impingers. The solution with trapped SO_2 gas is then transferred to culture tubes kept in thermostats, and sent to the chemical laboratory for analysis. At a time two sets of samples were collected simultaneously from two pits using two pump sets and one 12-volt battery.

5.2.5.2 Discussion of Results

Soil-gas samples of SO_2 collected in the field by running the pump for a period of ten hours from above the soil-covered sulphide bearing zones of Aladahalli and Chitradurga areas, Karnataka could detect 2 to 41 micrometres of SO_2 gas per 25 ml of above mentioned solution, while the samples from the pits away from the sulphide lode generally recorded N.D. to 5 microlitres per 25 ml (Talapatra et al., 1981). However, some of the samples collected from above the mineralized zone could not trap any SO_2 as per analytical results. These discrepancies may be either due to some drawbacks in the collecting mechanism or due to inadequacies in the sensitivity and specificity of the analytical method used, which requires further R & D efforts standardization of the analytical method.

5.2.5.3 Conclusion

From the above account it is clear that mercury-vapour trap sampling technique has been proved to be very useful for exploration of concealed sulphide deposits, especially in arid to semi-arid areas covered by transported sandy soil, alluvial/desertic sand, talus etc. It has been proved beyond dispute that once the flow of Hg-vapour gradient to the surface is established, the thickness of overburden or its character is not of much importance, provided it is porous and the degassing of mercury is

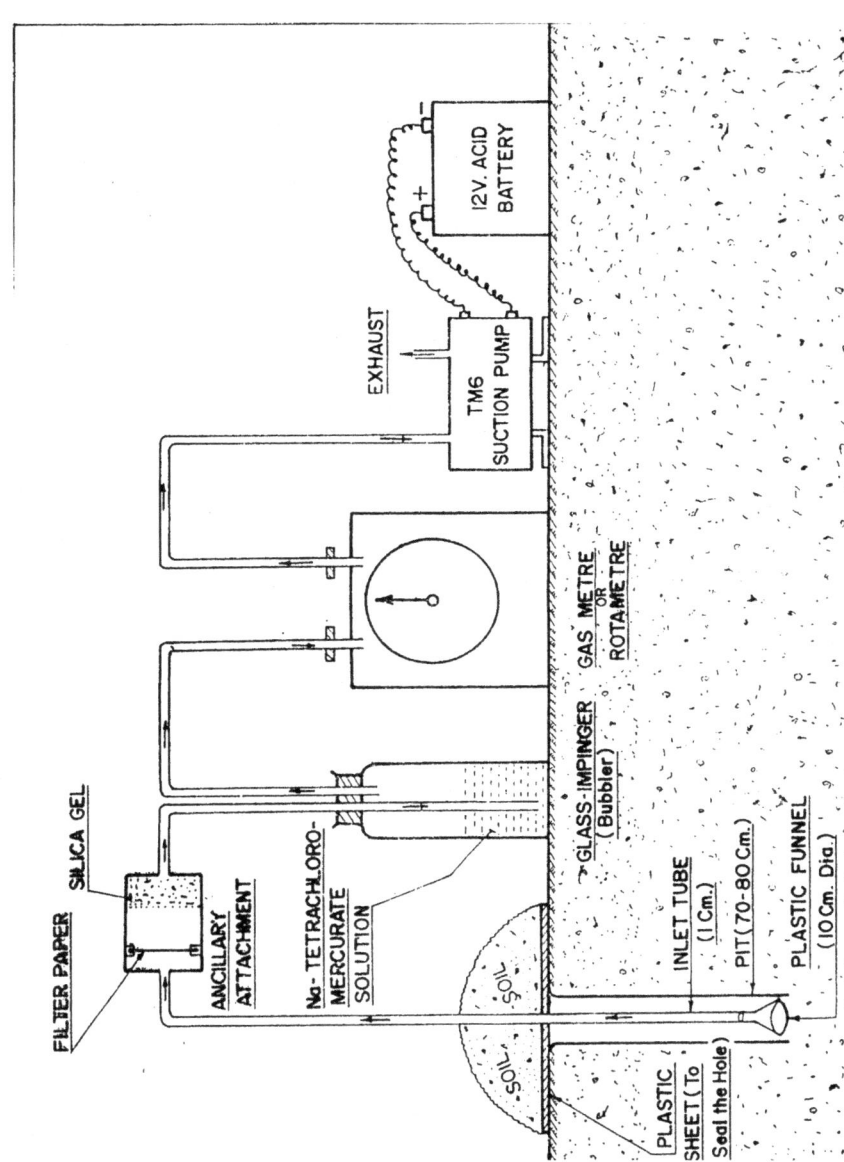

Fig. 5.5 Flow chart showing SO₂ soil-gas sampling method

continuous in that section (cf. McNerney and Buseek, 1973). This vapour trap sampling technique has been applied with success in parts of Rajasthan, Jharkhand and some other parts of Indian sub-continent by G.S.I.

Simplicity of this technique and its low cost of sampling are the added advantage over other vapour-phase geochemical methods, provided proper laboratory facilities are available for quick and precise analysis of Hg in the trap samples. However, gold instead of silver will be more suitable as a trapping material for repeated use. Cheaper substitute will be gold plated Ni-Cu-Cr wire alloy. The technique of Hg-vapour trap sampling, thus, rightly deserves wide application for exploration of concealed base metal sulphide deposits in areas with thick overburden present in different parts of Indian subcontinent after more R & D efforts in major parts of NW India. This will, however, require proper experimentation by repeated sampling in different periods of the year for standardization of the technique in respective areas.

As regards SO_2 soil-gas sampling technique, it appears that SO_2 soil-gas sampling requires more research and development work to bring it to the degree of sophistication at par with the other geochemical methods before its wide application in exploration. Since the nature of SO_2 anomaly depends on a number of variable factors mentioned earlier, interpretation of soil-gas data should be done with proper caution.

In India preliminary studies with soil-gas on these lines have been carried out in known mineralized belts with locally developed equipments as described earlier, and some techniques have been standardized for future work (Talapatra and Bose, 1979; Talapatra et al., 1981). Several other soil-gas surveys have proved worthwhile elsewhere in the world. It has been observed that organo-metallic compounds are susceptible to airborne surveys, while the CO_2 anomalies even in places with 70-80 m overburden can be detected. Helium and radon have been used for uranium exploration in Wyoming and Utah, and sulphur dioxide and carbon dioxide have been used to detect sulphide deposits under pediment gravels. Bromine and iodine in traces have also been found in the air over some copper porphyry deposits (McCarthy Jr., 1972). These non-conventional techniques of soil-gas sampling, specially for Hg and SO_2 may be adopted in any part of the world where concealed sulphide deposits are suspected. However, this requires proper experimentation by repeated sampling.

5.3   FIELD ELECTROCHEMICAL TECHNIQUE

This is another Research and Development (R&D) Project of GSI for discovering concealed base-metal sulphide deposits in any terrestrial areas that has been successfully developed by the earth-scientists of GSI which is expected to be fruitfully applied in any part of the world. A method based on the principles of electrochemistry has been applied for the first time with indigenous equipment in the search for base-metal sulphides, following a field exploration technique known as 'CHIM-10', which is in use in the U.S.S.R. for the last few decades. The technique utilizes the electro-positive property of the metallic ions. When a large section of the ground is artificially electrolyzed by the introduction of a direct current, the metallic ions within the moist soil profile close to the mineralized zone tend to move towards the energizing cathodes of a powerful widely spaced electrode array (Talukdar et al., 1985).

An attempt was made by Talukdar et al. (1985) to assemble a similar equipment in small scale as a prototype testing the technique in the laboratory so as to examine in detail the technique's effectiveness and applicability which subsequently was a grand success under the Indian conditions (Talapatra et al., 1986b). With a view to completing the job in a short time, some of the available geophysical instruments of the Department with minor modifications were utilized for fabrication of a prototype system. Trial runs were made across the mineralized zones of a copper sulphide deposit of Karnataka, India near Aladahalli. Similar technique was successfully carried out over the mineralized bodies of Singhbhum District, Jharkhand, India by Banerjee et al. (1992) where metallic mineral exploration by partial electrolytic extraction in the field was carried out.

5.3.1   Methodology

The metallic ions which are carried away by a direct current while passing through the moist soil medium with sulphide ore minerals can be trapped in a specially designed container with the cathode electrode immersed in an acidic medium as mentioned earlier. In a similar laboratory model, metallic ions are observed to have been transported through the moist soil by a strong direct current. The deposition of these metallic ions, both on the cathode rod and in the acidic medium (electrolyte), is found to be quite significant and within detectable limits when an electric current is passed continuously over a long period of time.

In the field condition also a similar type of partial extraction of metals from the weathered soil profile is possible by using a field technique utilizing the method of electrolysis where the moisture within the pore spaces acts as the electrolyte/carrier. The method requires passing a stable high-voltage direct current for a long period of time, say 15-20 hours, between an anode and an array of special type of cathodes widely spaced along a profile over sub-surface ore body. The electric current causes the positively charged metallic ions to move towards the cathodes under the influence of a potential field. The ions deposited on the cathode are identified and chemically analyzed. Thus it is possible to detect the metals associated with the deposit. It may also be possible to estimate approximately the reserve of the deposit by computing the collected ionic mass with time in known area provided proper R

& D work is carried out. The technique will thus help locate metallic ore bodies lying at depth on the basis of anomalous concentration of metals at the cathodes along the profile.

5.3.1.1 Laboratory Model Experiment

The method was initially assessed in the laboratory by conducting a quick preliminary experiment on a small model. It consisted of moist sandy soil placed in a large plastic tray (1.0 × 0.6 m) to form a layer about 20 cm in thickness. A rich sample of polymetallic sulphide were pulverized in to fine grains and placed underneath this soil layer at one end of the plastic tray. A small steel rod was used as anode, which was placed in the soil on the other end of this tray. One small pyrex-glass crucible with a fine porous plate at the bottom was located on the other end of the tray within the soil to form the cathode assembly. The crucible was filled with a 2N solution of HNO_3 in which a small graphite rod (4 mm diameter) was inserted duly secured by a clamp. A Hewlett-Packard DC stabilized power supply was used as a constant current source. A current of 80 mA was placed for a period of three hours. The electrolyte and the graphite rod were then separately analyzed for metal values that have been deposited in these.

5.3.1.2 Results

During the model test, two sets of data were collected under the same laboratory conditions on different days, for two different types of pulverized polymetallic sulphide feed placed under the soil cover below the cathode positions. Test runs were carried out with 60 ml of 2N nitric acid placed in the crucible each time. The acid showed reduction in volume after each 3-hour run due to seepage into the soil and evaporation which was equally replenished at regular intervals. Analytical results of copper, zinc and lead in microgram determined by AAS are given in Table 5.2, indicating very interesting results.

It is observed that both electrolyte and graphite rod samples trapped considerable amounts of copper, lead and zinc, and more of these metals were trapped when relatively richer polymetallic sulphide ore sample (approx. 75% total sulphide) was used as feed.

Table 5.2 Results of laboratory tests

Nature of sulphide used	Volume (ml)	Electrolyte			Graphite Rod		
		Cu (mg)	Zn (mg)	Pb (mg)	Cu (mg)	Zn (mg)	Pb (mg)
Poor ore	38	1.1	0.8	_	46.0	47.0	7.0
Rich ore	39	25.0	2888.0	76.0	830.0	470.0	27.0

5.3.2 Geology of the Aladahalli Belt

The field sampling of this method was standardised in the Aladahalli mineralised belt of Karnataka where GSI has explored the area by drilling. The base-metal mineralization of the Aladahalli belt of Karnataka (vide inset map, Fig. 5.6) is associated with the metamorphic rocks of the Precambrian age. The sulphide mineralization

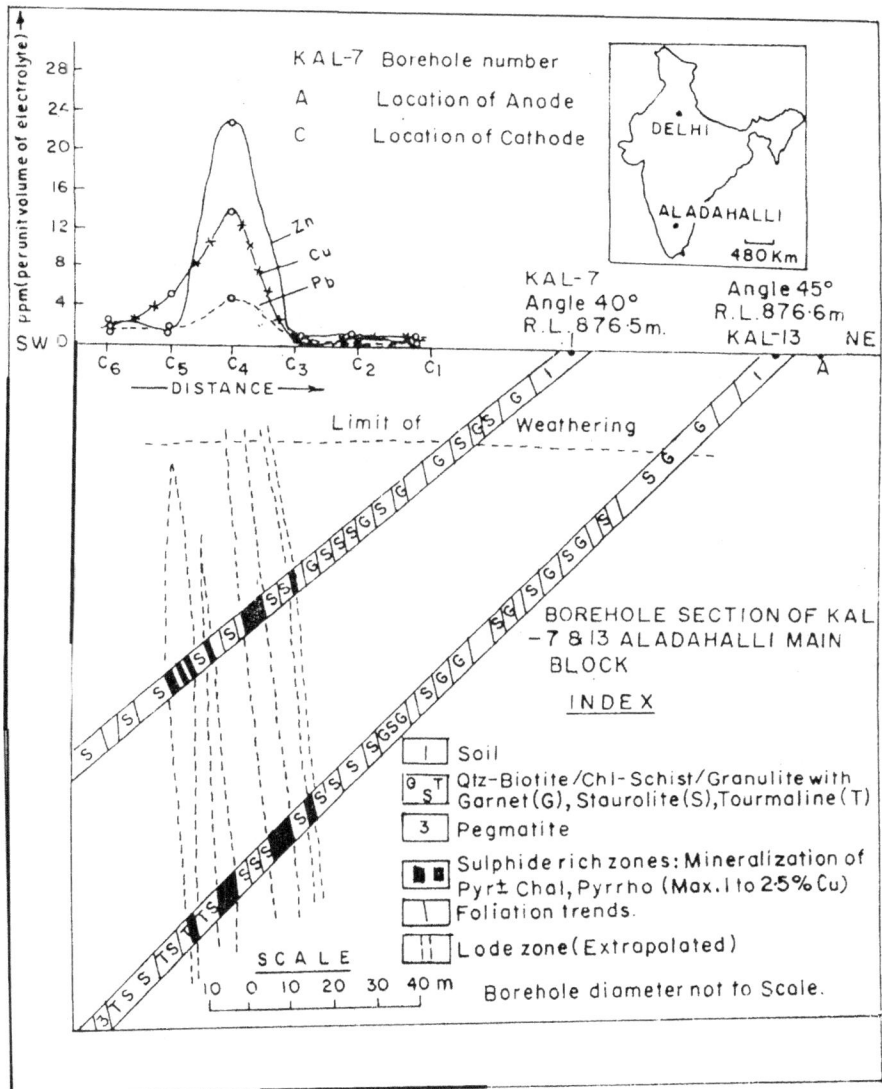

Fig. 5.6 Results of field sampling by electrochemical method

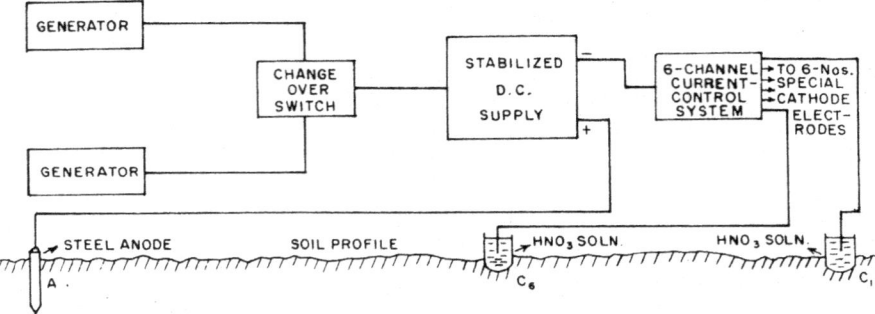

Fig. 5.7 Flow chart showing field electrochemical sampling

of this area is confined in a NNW-SSE-trending narrow schist belt (Jayaram and Ravindran, 1979) occurring within the Peninsular Gneissic Complex. Its manifestations on the surface are indicated by a number of parallel and discontinuous bands of gossan occurring in the undulating ground. The gossans exhibit shades of brown, brick-red, maroon, black, yellow and orange colours. Disseminated sulphides are seen to occur in almost all the rock types of the area, but these are concentrated in chloritic mica schists and brecciated rocks containing staurolite, garnet, muscovite and occasionally tourmaline. The schistose rocks are either vertical or steeply dipping towards NE and the lode zones are generally conformable with the schistosity. The sulphides comprise predominantly pyrite with some chalcopyrite and pyrrhotite in varying proportion. A trial run of the electrochemical technique in the field was conducted on a soil covered flat to gently sloping area along a line trending NE-SW and passing over two boreholes drilled by GSI in the Aladahalli Main Block, where rich sulphide zones were intersected within quartz biotite/quartz-chlorite schists (Fig. 5.6), with occasional pegmatite veins.

5.3.3   Field Sampling Method

The prototype equipment used in the field consisted of two ordinary motor generators (including one as a stand-by) for supplying continuous power to the stabilized D.C. supply, a 6-channel current control unit, six special types of cathode electrodes and one single anode (Fig. 5.7). Motor generators used were rated at 5 HP, 110 V, 400 Hz, 2.5 kW. In fact, the two motor generators of equal capacity were employed using one at a time with a change-over switch arrangement. This was done in view of the requirements of a continuous supply of power over a long period of time in case of failure of one of the units. An IP Transmitter (frequency domain) manufactured by Geoscience, USA (now no longer available) was utilized as a stabilized current source of up to 5 A, after incorporating the necessary modification in the electronic circuitry to provide the direct current. The 6-channel current control unit was devised to facilitate independent monitoring of current in six different cathodes each through a current meter. Each of the special cathodes consisted of a porous ceramic pot with an unglazed surface at the bottom. Each pot contained one graphite rod held at the top by a Perspex disc. The pots containing a graphite rod, each immersed in an equal quantity of acid, formed the special cathodes. A screw arrangement in the disc ensured perfect electrical contact with the graphite rod. On the other hand, the single anode consisted of a number of sturdy steel electrodes. The alternating current supplied by the motor was converted into a very stable direct current source. The 6-channel current control unit in turn allowed the same amount of current to pass to each of the six cathodes.

A measurement traverse was made along the line joining boreholes KAL-13 and KAL-7 (Fig. 5.6), after Ravindran (1982). The anode was placed on the hanging wall side of the lode zones near KAL-13 at a distance of about 100 m from the nearest cathode. The cathodes were spaced 15 m apart towards southwest. The current was maintained at 165 mA per channel. Total duration of current flow during the trial was 15 hours. Initially 300 ml of electrolyte (4N nitric acid) was put in each cathode. As the process continued, equal quantity of electrolyte was added to each cathode pot from time to time to make up the loss of electrolyte due to seepage into the ground through the porous bottoms of the cathodes and evaporation from their surfaces. At the end of the run, the graphite rods and the remaining electrolytes were taken out separately from the cathode assembly for analysis. Blank samples of electrolyte and graphite rod were also kept separately in order to determine the background level of contamination, if any. All samples were analyzed in the chemical laboratory of AMSE Wing, Bangalore.

5.3.3.1   Discussion of Results

The analytical results of field samples showed anomalous concentration of zinc, copper and lead values per unit volume of electrolyte with distinct peaks above the mineralized zones (Fig. 5.6), which were about 20 times, eight times and three times that of respective background values. However, no significant concentration of metal could be detected within the blank samples kept in a cathode away from the line of sample. It is interesting to note that the highest peak is shown by zinc values followed by copper and lead, lead being least mobile with respect to the other two elements. It may be mentioned here that the sample from brownish clayey 'B' horizon soil collected from a pit about 40 cm in depth located above the lode zone close to the referred traverse line recorded 600 ppm Cu and 100 ppm Zn, while an *in situ* gossan sample from near the same mineralized zone lying 50 m to the NW contained 2000 ppm Cu and 300 ppm Zn.

The field electrochemical method standardized in the Aladahalli mineralized belt, Karnataka, thus, appears to have picked up distinct anomaly above the mineralized zone. It is, therefore, surmised that the method may be quite effective in locating concealed base-metal sulphide zones where the load zone is covered in any area. Similar electrolytic extraction of metals was effectively utilized for detection of

concealed polymetallic minerals from Singbhum District, Jharkhand by partial analyzing of the electrolyte chemically (Banerjee et al., 1992). However, more detailed
work in diverse field conditions with variable time of operation and other physico-
chemical parameters requires to be done as prerequisite to judge its applicability in
different types of mineralized belts. This technique may be applied along any mineralized belt with concealed deposits in any part of the world.

5.4   RADIOGENIC ISOTOPES AND STABLE ISOTOPES IN EXPLORATION GEOCHEMISTRY

The study of radiogenic isotopes and stable isotopes may be very useful in mineral
exploration efforts specially for sulphide deposits which will be discussed in the
following paragraphs. It may be mentioned here that the time and environment of
mineral deposits have a pattern on the global scale. The radiogenic isotopes could
be used to define the time of formation and the stable isotopes to indicate the environment of its formation. Thus an effort to use correlated stable and radiogenic
isotopic studies will be beneficial to understand the perspective of the formation of
mineral deposit.

5.4.1   Study of Radiogenic Isotopes

The role of radiogenic isotopes pertaining to Rb-Sr, Sm-Nd and K-Ar systems apart
from Pb isotopes help in timing the formation, in identifying special signatures
indicating their locale and help in an assessment of their commercial viability. The
role of initial ratios is quite evident in assigning proper postulates regarding their
geologic origin. Their role in exploration is *mostly indirect*.

The most important derivatives of radiogenic isotopes, which are of use, pertain
to the daughter Pb isotope derived by the decay of radiogenic uranium and thorium
isotopes. Lead has four isotopes of which 206, 207 and 208 are derived as follows:

$$^{238}\text{U} \rightarrow {}^{206}\text{Pb}$$

$$^{235}\text{U} \rightarrow {}^{207}\text{Pb}$$

$$^{232}\text{Th} \rightarrow {}^{208}\text{Pb}$$

^{204}Pb is non-radiogenic in origin. By utilizing the ratio of radiogenic lead to that
of non-radiogenic lead, several useful parameters have been evolved. The study of
lead isotopes has been utilized in not only evolving a model lead age but also in
identifying the possible source rocks and evaluation of ore genesis, i.e.,

magmatogenic, indirect magmatogenic, metamorphic, lateral secretion, syngenetic, vein deposits of uncertain origin, ores possibly related to brines and in prospecting for old uranium deposits (Doe and Stacey, 1974). It is proposed to highlight some aspects only which are directly relevant to exploration.

The major base metal deposits have characteristic lead isotope ratios (Gulson, 1977) and may have followed a single stage evolution whereas minor deposits indicate that lead has not evolved under such conditions. An economic deposit may be indicated if the $^{208}Pb/^{204}Pb$ model age is in agreement with primary isochron age. The probability of a prospect turning out to be a major mine may be predicted by a statistical analysis based diagram using data from producing major mine. In case the isotopes ratio of a prospect is similar to that of a producing mine in the same mining district, the chances are that this prospect may also become a major producing mine. The indication of uraniferous lead in galena from a mineral prospect implies a very high probability that this particular galena is associated with uranium mineralization. An analysis of ordinary lead of about zero model age may convey hardly any information about a mineral occurrence except that the deposit is almost certainly of Phanerozoic age. In the North American continent another variety of ordinary lead, with model age of 2800 Ma, interested gold hunters whereas that with model age of 1600 Ma interested base metal prospectors.

5.4.2 Study of Stable Isotopes

Stable isotope studies have been widely used in the study of ore deposits (Ohmoto, 1986). Variation of light stable isotopes are useful, when used together with other geochemical, textural and field data; these pertain to H, C, N, O, Si and S. Two notations are used in the stable isotope studies: (a) the d (delta) value which represents the difference in absolute isotopic ratios between the sample and a standard, and (b) the fractionation factor, which is roughly the difference of the value between the two samples. The fractionation factor varies as $1/T^2$ and is extensively used for geothermometry.

5.4.2.1 Isotopic Variations in Nature

In oxygen isotopic system, evaporation of water enriches the water vapour in the lighter isotope. Thus snow and rain water are isotopically lighter than the ocean water. The interaction between meteoritic as well as the water with rocks and sediments causes isotopic variations in nature. Such isotopic variations are temperature dependent. Sulphur isotopic variations, on the other hand, depend on the radox reaction, which can be biogenic or non-biogenic, involving different isotopes of sulphur. In general, components with higher oxidation state are isotopically heavier. Thus

$$\Delta 34\,S \quad > \quad \Delta 34\,S \quad > \quad \Delta 34\,S$$
$$(\text{sulphates}) \qquad SO_2 \qquad (\text{sulphides})$$

5.4.2.2   Oxygen Isotopes

Oxygen isotopic studies are particularly useful in deciphering the hydrological aspects of ore forming systems, e.g. the origin and the mass of the ore-forming fluids, geometry of the plumbing systems, duration of hydrothermal systems, etc. (Green et al., 1983). Since water is the medium for all ore constituents (except in orthomagmetic deposits), study of the 'O' isotopes lead to the understanding of the genesis and evolution of such ore bearing fluids. Because of the high value of the fractionation factor at low temperatures, interaction with meteoric water at low temperatures causes an isotopic enrichment in the altered rocks. On the other hand, high temperature interactions, typical of hydrothermal fluids, cause depletion in the oxygen isotopic ratio of rocks. A meteoric hydrothermal system accompanies many granitic intrusions and causes concentric aureoles of increasing d18 O away from the pluton. In the low d18 O zone many ore deposits are found.

5.4.2.3   Sulphur Isotopes

The most important reservoir of global sulphur is the ocean water which contains on an average 900 ppm of sulphur and has a delta 34 S value of +20%. In contrast, mantle rocks have delta 34 S varying from –3 to 2%. While the sedimentary rocks contain on an average 5000 ppm of sulphur, igneous and high grade metamorphic rocks contain about 200 ppm of sulphur. Sulphur is removed from the sea water as sulphates (gypsum) or sulphides in the ratio 1:2. It is obvious that the sea water plays an important role in the genesis of sulphide ores. The most important sulphur isotopic fractionation takes place by the reduction of marine sulphates by sulphur reducing bacteria *Desulfovibrio desulfuricans*. The resultant sulphide is enriched by 50% in d32s. In contrast, sulphur in the form of SO_4 in evaporates show similar isotopic characteristics as the sea water. If the evaporite, SO_4 is subsequently inorganically reduced, the resulting sulphide will have isotopic characteristic similar to the original SO_4 (Ohmoto and Rai, 1979).

5.4.3   Significance of Stable Isotope Studies in Sulphide Deposits

5.4.3.1   Porphyry and Skarn Type Deposits

According to the nature of the associated granite pluton, porphyry and skarn deposits can be devided into 'S' and 'I' types, While d18 O values show characteristic mantle ('I' type) and sedimentary ('S' type) values, the delta 34 S from both type of deposits show wide and overlapping values. This suggests that the ore forming potential of the granitic plutons depends on the upper crustal cover rocks with which these plutons interact and not as such on the source region of the granitic plutons.

5.4.3.2   Vein and Replacement Type Deposits

Mineral deposits containing Zn ± Pb ± Au ± Ag ± W that form in continental settings occur mostly in veins and as disseminated mineralisations in volcanic and sedimentary rocks or as replacements in limestones. Isotopic studies indicate that most of this group of ores form at temperatures between 150 °C and 350 °C and that the ore carrying fluids depositing Pb-Zn-Cu are more saline than 'Au only' deposits. Such studies in conjunction with conventional geothermometry and fluid inclusion studies can be useful in formulating the strategy for exploration.

5.4.3.3   Massive Sulphide Deposits

Massive Pb-Zn-Cu-Fe sulphide are found associated with submarine volcanic rocks near the ridge crest. Oxygen isotope geothermometry indicates 150 °C-350 °C temperature of formation and H, O and S isotopes indicate that the ore forming fluids acquired sulphur by two different processes. Some of the sulphides were acquired from the wall rocks where they were first incorporated as SO_4, and were then reduced by Fe and C. The second process was bacteriogenic reduction of sea water-sulphate. Disequilibrium of isotopic fractionation between the coexisting galena and sphalerite have been interpreted as indicative of mechanical disaggregation, transportation and redeposition of ores of different generations. Isotopic disequilibrium between chalcopyrite and other sulphides support the model of formation of Cu ores by replacement of earlier Pb-Zn ores.

Stable isotopic studies help in providing meaningful answer to mainly geological problems when used in conjunction with radiogenic isotopic studies and related field and laboratory data. As applied to the problem of ore genesis, stable isotopic studies have demonstrated the complex interplay of various near surface processes in forming the ore deposits. A knowledge of the temperature of formation of ores, and the genesis of ore bearing fluid can be obtained with the stable isotope studies. Such studies may be useful in mineral exploration efforts.

Chapter 6
APPLICATION OF MODELLING AND GEOCHEMICAL TECHNIQUES FOR REE AND RARE METAL EXPLORATION

6.1 INTRODUCTION

Rare Earth Elements (REE) and Rare Metals (RM) have become an essential commodity for modern living in recent times. As such exploration of ore minerals of these elements requires special attention (Goodenough et al., 2017). However, geostatistical method for deposit modelling along with geochemical techniques has become a common practice in recent years for evaluation of mineral resources and for the possible prediction of the resources in the study area (Agterberg et al., 1972; Benest and Winter, 1984; Bonham-Carter et al., 1989; Talapatra, 2006), as discussed in Chapters 1 and 2. These methods may be applied for the undiscovered mineral resources of a region including REE and RM, especially in extension areas of known mineral belts and also in virgin areas where characteristic interrelationship of different geological, geochemical and geophysical variables can be established. The Atomic Minerals Directorate (AMD) of Government of India is running one of the country's leading multidisciplined and multifaceted exploratory organization since 1950s, especially for uranium, thorium, Rare Metals (Nb, Ta, Be, Li etc.), Rare Earths (La, Lu etc.) and yttrium. It has been playing a pivotal front-end role in the country's atomic energy and similar scientific programme through exploration of related mineral deposits, required for nuclear and hi-tech industries like electronics, telecommunications, information technology, space and defence etc.

The targets for exploration for the REE and RM mineral resources are essentially the primary pegmatite sources as well as the coastal placer and elluvial concentrations derived from them. Initially qualitative geological variables from coarse grained pegmatities and related rocks of Purulia-Bankura belt, West Bengal have been examined in the light of earlier work done by Talapatra et al. (1986a, 1991). Subsequently, detailed geochemical studies of the collected stray samples from the

© Capital Publishing Company, New Delhi, India 2020
A. K. Talapatra, *Geochemical Exploration and Modelling of Concealed Mineral Deposits*, https://doi.org/10.1007/978-3-030-48756-0_6

area give sufficient clue for detecting occurrence of concealed deposits of rare earths along the area studied (Talapatra, 2004, 2014). In this respect, search for Rare Earth Elements and Rare Metals of any country may be utilized with the help of these techniques described in this chapter. Rare earth elements (REE) are generally found in Precambrian rocks namely pegmatite, carbonatite etc. along crustal areas of earth (Bhadra et al., 1989) and also within the beach sands mostly along coastal areas where heavy minerals like monazite etc. are commonly found.

In this regard, Rare Earth deposits may be divided into two principal types, namely andogenic and exogenic. The former includes some carbonatites in association with pegmatites, metamorphic-metasomatic veins, and the latter comprises coastal or beach placers, inland placer and offshore placers. Among the principal ore minerals of REE in Indian exogenic type of deposits is monazite, although xenotyme holds out some prospect for the future. Out of India's estimated reserve of five million tonnes of monazite, 70-75% are in beach placers and the rest in the inland and offshore varieties (Sarkar et al., 1995a, b). In the development of Indian beach placer deposits, granites, granitic pegmatites, migmatites, gneisses, charnockites, leptinites and khondalites generally act as source rocks, and the tropical climate with heavy rainfall and strong wave action along sea coast was specially conducive to the concentration of the placer minerals.

6.1.1 What Are Rare Earth Elements

Rare earth elements are a family of 17 elements in the periodic table, namely scandium, yttrium and the 15 elements of lanthanide series. The seventeen rare earth elements with symbol and their essential uses are listed below:

1. Scandium (Sc) – Aluminum alloy in aerospace, 2. Yttrium (Y) – Phosphorous, ceramics, lasers, 3. Lanthanum (La) – Re-chargeable batteries, 4. Cerium (Ce) – Batteries, catalysts, glass polishing, 5. Praseodymium (Pr) – Magnets, glass colorant, 6. Neodymium (Nd) – Magnets, lasers, glass, 7. Promethium (Pm) – Nuclear batteries, 8. Samarium (Sm) – Magnets, laser, lighting, 9. Europium (Eu) – TV colour phosphorous: red, 10. Gadolinium (Gd) – Superconductors, magnets, 11. Terbium (Tb) – Phosphorous: green, florescent lights, 12. Dysprosium (Dy) – Magnets, lasers, 13. Holmium (Ho) – Lasers, 14. Erbium (Er) – Lasers, vanadium steel, 15. Thulium (Tm) – X-ray source, ceramics, 16. Yterrbium (Yb) – Infrared lasers, high reactive glass, 17. Lutetium (Lu) – Catalyst, PET scanners.

It can be seen from the above statement that all these elements are used in one or more hi-tech products necessary for modern living. Significant rare earth minerals include ilmenite, sillimanite, zircon, monazite etc. commonly in association with beach sands and sometimes with pegmatites, carbonatite etc. in Precambrian crustal areas. In this respect REE should be studied in details (Henderson, 2013; Massari and Ruberti, 2013; Nassar et al., 2015; Roskill, 2016) for exploration of these mineral deposits.

The rare earth elements (REE) mentioned above include the lanthanides, from lanthanum (La) to lutetium (Lu) and often include the chemically similar elements like yttrium (Y) and scandium (Sc). They are typically divided into two groups, the light and heavy rare earth elements (LREE and HREE, respectively), with LREE including lanthanum (La), cerium (Ce), praseodymium (Pr), neodymium (Nd) and samarium (Sm). The HREE then extend from europium (Eu) to Lutetium (Lu). Yttrium is grouped with the HREE due to its similar properties. The REE are widely viewed as critical metals, because they are extensively used in modern technology.

In the earth's crust, there is a general trend of decreasing abundance of REE with increasing atomic number. Accordingly elements with an even atomic number are more abundant than those with odd atomic numbers. This means that Ce is the most abundant of the REE in earth's crust; in contrast, Lu is genuinely rare. Most natural REE ores are dominated by La, Ce and Nd with much smaller amounts of HREE.

Common REE and RM bearing minerals present in earth may be *carbonate* namely, bastnaesite (Ce, La) FCO_3 with larger content of Eu, *phosphate* like monazite (Ce, La, Th, U) PO_4 and xenotime (YPO_4), *oxide* like loparite (Ce, La, Ca, Na) $(Ti, Nb)_2O_6$ and lastly silicate like gadolinite $(Be_2FeY_2Si_2)_{10}$.

6.1.2 Major Rare Earth Producing Countries of the World

Here is a list of eight countries that mined the most rare earths in the year 2017, as per the United States Geological Survey, but US missed out on rare earths production during that period. However, most recent data believe that the demand for these metals is set to rise and in fact some predict that the market will be valued at $20 billion by 2024. **China** has dominated rare earth production for a number of years. In 2017, its output of 105,000 MT was unchanged from the previous year and it had exported 39,800 tonnes of rare-earth materials, a 10% increase compared with exports from the same time period in 2016. Though China is the world's largest supplier of rare earths, it is surmised that it may put an annual limit on its rare earths production from 2020. Next to China is **Australia** whose rare earths production is around 20,000 MT, compared to 15,000 MT in 2015. However, rare earths have only been mined in the country since 2007, though it holds huge reserves of rare earths. Next to mention is **Russia** whose rare earths output in the year 2017 was 3000 MT, that increased from 2800 MT from the prior year. Despite the slight increase in production, Russia will increase over time through the development of pre-existing rare earths fields. Back in 2012, an $8.4 billion rare earths deposit was discovered in **Brazil.** However, in the year 2017 only 2000 MT was produced, while prior to this year production was 2200 MT. It seems little ore production is there compared to the deposit discovered. Next country in order of rare earths production is **Thailand**, which produced 1600 MT in both 2016 and 2017. Its total rare earths reserves are not clearly known, but the country remains a fairly significant producer outside China.

Coming to rare earths reserve of **India**, the country holds almost 35 percent of world's total beach sand mineral deposits, which are significant sources of rare earths due to presence of heavy mineral, namely monazite. Figure 1.3 of Chapter 1 shows the most important types of heavy minerals present along the coastal areas of India. India's current rare earths production is far below its potential, which is about 1500 MT in the year 2017, unchanged from the previous year. Indian Rare Earths Co. Ltd. and Toyota Tsusho Exploration entered into an agreement regarding the exploration and production of Indian rare earth deposits via deep sea mining in recent years. Besides the beach sand deposits, pegmatites in association with carbonatite bodies along prominent lineaments of Precambrian terrains of Indian subcontinent are expected to unearth significant deposits of rare earths in future (Fig. 6.1). Techniques of discovering such deposits from beach sands as well as along cratonic blocks of Precambrian shield areas of India have been discussed in this chapter with special reference to deposits of Purulia-Bankura area, West Bengal.

Coming to **Malaysia's** production of rare earths of about 300 MT in 2017 was unchanged from the previous year. The country is home to one of the world's largest rare earths refineries, thereby becoming an important player in the rare earth's space. It sells its products to Japan, Europe, China and North America. Coming to the eighth country, namely **Vietnam** which decreased its rare earths production from 220 MT in 2016 to 100 MT only in 2017, indicates that mining of rare earths in Vietnam from its ore reserve is also scarce in recent years.

Global distribution of the REE is very uneven, with the proven reserves largely distributed in China (43%), the Commonwealth of Independent States (19%), Australia (5%) and the rest from the remaining countries. The CIS i.e. Commonwealth of Independent States includes Azerbaijan, Armenia, Belarus, Georgia, Kazakhstan, Kyrgyzstan, Moldova, Russia, Tajikistan, Turkmenistan, Uzbekistan and Ukraine. Most of the global REE production, however, is in China, with estimates ranging from 95 to 97 percent. About half century ago China was not among the leading producers of REEs. Between 1950 and 1980, the U.S.A., India, South Africa and Brazil were considered to be the front-runners in production of REEs. During the 1980s, China began underselling competitors, leading to consumers purchasing cheap supply from the Chinese land. Of late, significant amounts of REEs are produced in only a few countries and China is the dominant producer of it now. Other countries with notable production in 2009 were Brazil, India, Kyrgyzstan and Malaysia. Minor production may have occurred in Indonesia, Nigeria, Korea and Vietnam.

6.1.3 Rare Earth Elements—An Essential for Modern Living

Before 2010, hardly any one even knew what are the use of rare earths. Twenty years ago there were very few cell phones in use by common people, but the number has risen to more than five billion or so today. Computers and DVDs have grown almost as fast as cell phones. The rechargeable batteries, used in portable electronic

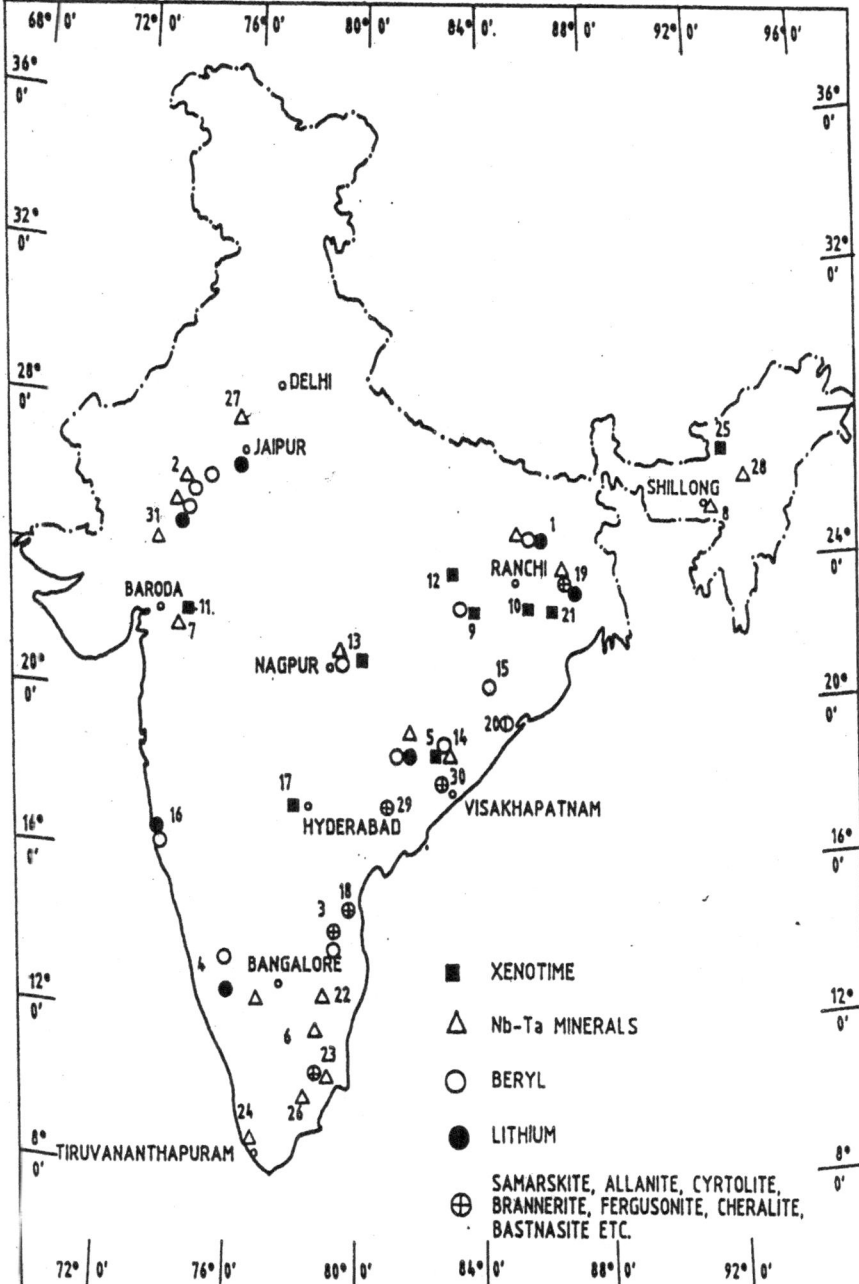

Fig. 6.1 Rare Metal and Rare Earth pegmatites occurring within Indian sub-continent after Banerjee (1999) located in Arunachal, Assam, Meghalaya, Bihar, Rajasthan, Gujrat, Maharastra, Madhya Pradesh, W. Bengal, Orissa, Telegana, Andhra Pradesh, Karnataka, Tamilnadu and Kerala

devices such as cell phones, computers and cameras are made with rare earth compounds. Besides several kilograms of rare earth compounds are used in batteries that power electric vehicles. As concerns for energy independence, climate change and other issues, the sale of electric vehicles will increase which obviously demand batteries made with rare earth compounds. Rare earths are also used for air pollution, phosphors and polishing compounds. These are also used for air pollution control, illuminated screens on electronic devices and optical quality glass. Rare earth elements play an essential role in defence too. Rare earth metals are also key ingredients making the very hard alloys used to make armoured vehicles and projectiles that shatter upon impact in thousands of sharp fragments.

The REEs have a unique arrangement of electrons that lends special properties to materials containing these. Because of these, REE are in great demands for high tech materials. Among these properties is one that lends incredible magnetic strength to magnets made with REEs. These rare earth containing super-magnets are utilized in computers, automobiles and other vehicles (including hybrid vehicles), consumer electronic products, medical products, systems and motors of all kinds. They also add functionality to jet fighter engines, electronic counter measure systems, missile systems and satellite communication systems.

6.1.4   Changing Demand for the Rare Earth Elements

Since the initial discovery and extraction, the use of REE have gradually changed to high-priority rare earth metals that are being used in advanced electronics, lighting, power generation and military applications. In the developed world, REE are integral to many industrial, commercial and residential appliances and in the increasing electrification of vehicles. Specially, the use of REE in hybrid electric vehicles (HEVs) and full electric vehicles (EVs) are significant, which are expected to cause wholesale changes in the volumes and types of raw materials consumed by the automobile industry.

This production growth is likely to change greatly the demand for neodymium-iron-boron (NdFeB) magnets, which are already used in automobile industry. Most internal combustion engine (ICE) vehicles which use HEVs and EVs require small electric motors in components such as widescreen wiper motors and air conditioning system. However, the ongoing electrification of vehicles has increased the use of NdFeB magnets used in the power train in HEVs and EVs as well as in numerous other applications.

Recently wind turbines are second largest end-use applications for NdFeB magnets. China is the largest wind power generator, increasing its installed capacity by over 70 GW between 2013 and 2015. The direct drive permanent magnet generator (DDPMG) for wind turbines require largest volumes of NdFeB magnets. Newer DDPMG technologies have become commercially available for large-size turbines, used mainly in offshore wind forms which are expected to increase the volume of NdFeB magnets consumed per unit.

Rare earths are often used as additives in glass fibre optic applications to improve data transfer speeds. Erbium (Er) is the most commonly used dopant (varnish) with ytterbium (Yb) used in larger scale optical fibres. Other rare earths including thulium (Tm) and praseodymium (Pm) have also been used in some optic fibres though these are not commercially produced.

Production of two lightest rare earths, namely La and Ce is supposed to increase, depending upon their demand in next 10 years. Prices of La and Ce are likely to remain low. Despite this over supply, these elements are both critical to future generations, either in existing widespread technologies or technologies under development.

Lanthanum is used as an additive to improve the refractive index, decrease dispersion and enhance chemical stability in La-series optical lenses, mainly used in wide-angle lenses for consumer electronics. Yttrium (Y) and gadolinium (Gd) may be used as a replacement for lanthanum, though cost makes them unattractive substitutes. The increase in production of smart phones and tablets containing optical lenses has deserved a surge in demand for La-series glass production. The development of cerium-iron-boron (CeFeB) magnets in China has been found as a low cost alternative to NdFeB magnet with less supply availability risk.

The use of rare earths in phosphors has been in decline since the development and commercial production of LED technologies over florescent lamps. These lamps use Y as a host material and Lu, La and Ce as dopant material, while LED technologies use Y and Lu as the host materials, particularly in glow lighting. Europian (Eu) and terbium (Tb) are still used as small volume dopants in LED lighting products. LEDs are expected to continue replacing fluorescent lamps thereby reducing the demand for Y, Eu and Tb in phosphors, though warm-glass lighting has become more popular in commercial and residential lighting and demand for Lu is supposed to increase. The beneficiation of rare earth bearing minerals is most important in recent time (Jordens et al. 2013; Krishnamurthy and Gupta, 2015).

6.1.5 Rare Earth Deposits with Special Reference to Indian Scenario

In India rare earths are found in abundance within monazite etc. mineral present along the coastal sand deposits. A government panel is currently working on a strategy to give impetus to exploration and discovery of rare earth and energy critical elements used in renewable energy. The Atomic Minerals Directorate of India along with Geological Survey of India carried out detailed study for exploration and research on REE and has located huge deposits of rare earths minerals within the beach sand deposit along east and west coasts of India. Besides these, Precambrian pegmatites and carbonatite bearing rocks along crustal areas of sub-continent along the different provinces of India have indication of the presence of rare earths mostly as concealed deposits as shown in different parts of India (vide Fig. 6.1). It appears from this figure that right from Eastern India, namely, Samhampi in Assam, West Kameng in Arunachal Pradesh, Halwai and Kanyaluka from Bihar Mica Belt presence of REE are reported from continental parts of India (Banerjee, 1999). Presence of such minerals are also reported from Rajasthan Mica Belt, Ambadongar, Vaswa Nadi and Umedpur of Gujarat, Ratnagiri and Salai-Pannai of Maharastra, Koraput, Jharsuguda and Gokhandi of Odisha and in parts of West Bengal, Madhya Pradesh, Andhra Pradesh, Tamil Nadu, Karnataka and also in certain continental parts of Kerala. Bramhagiri in Puri district and Ganjam district of Odisha along the east coast of India are also notable for presence of huge placer deposits of Rare Earths as heavy mineral deposits.

Along these coastal stretch, heavy minerals such as ilmenite, rutile, zircon, sillimanite, garnet and monazite are separated for industrial use. Currently, Indian Rare Earths Co. Ltd. is setting up a processing plant in Chhatrapur in Odisha with capacity to produce 11,000 tonnes rare earth chloride. Similarly, along the west coast also huge amount of beach sands are present along Kerala, Karnataka and Maharastra where rare earth bearing monazite is also present within the beach sand, which has already been discussed in earlier chapters. The best known among these are Chavara in Kerala and Manaralakurichu in Tamilnadu. In November 2012, Japan signed a memorandum of understanding with India to enable the import of rare earths from India. The move will help Tokyo reduce its heavy reliance on China for the key resources used to manufacture high-tech products. With rare earth production at throttle, India could supply around 4100 tonnes a year, equivalent to roughly 10 percent of Japan's peak annual demand, the report said. The production and export will be conducted by a joint venture between Toyota Tsusho Corp. and state-run Indian Rare Earths Co. Ltd. Japan imported around 90% of its rare earth supplies from China during the previous year, and is hoping to reduce its dependence because of the risk that Beijing might curb exports. In this regard mention may be made about techniques of estimating the total resources of the heavy mineral concentration along the placer deposits of coastal areas of India as detailed in Chapter 1, using geostatistical modelling along coastal and offshore areas upto the Exclusive Economic Zone (EEZ) of India. In the present chapter, Precambrian crustal rocks in association with pegmatites and carbonatites present along the cratonic blocks of India, which are commonly associated with rare earth and rare metals, will be discussed in details.

6.2   RARE EARTH ELEMENT AND RARE METALS IN PRECAMBRIAN CRUSTAL AREAS

A number of minor base-metal and other occurrences of Eastern India are lying scattered along Purulia-Bankura shear zone area of West Bengal. Detailed surface and sub-surface exploration to prove the extension or presence of the concealed lode zones both along strike and dip direction in the area calls for huge investment of money, manpower and time. Preliminary resource prediction for the different mineral occurrences of this belt can, however, be made easily by applying computer aided modern statistical techniques, discussed earlier in Chapter 2, using the characteristic interrelationship of the different geological, geochemical and geophysical variables of even qualitative nature, where quantitative data are not always available.

Bed rock geochemistry of samples collected from northern part of Purulia District, West Bengal indicates exceptionally high rare earth element bearing mineralization within pegmatitic rocks along a roughly E-W trending mega-lineament passing through Beku (23°26′14″: 85°57′12″), lying further north, which contains some lensoid pegmatites with high rare metal concentration along with some alkaline carbonatite complex bodies. The area appears to be an ideal site of lineament controlled metallogenesis as a result of the Precambrian crustal evolution in this part of the Indian Shield. Detailed study of these fertile pagmatites described from the area is expected to unearth sizeable concealed deposits of rare-earths and rare metals which are very useful as raw materials in high technology and other mineral-based industries.

Before discussing the details of the specific work carried out in Purulia-Bankura area, West Bengal, mention may be made about the occurrence of REE in pegmatite-carbonatite and metamorphic-metasomatic veins in Indian perspective. Some of the Indian carbonatite complexes, particularly those of Ambadonger in Gujarat and Samalpatti and Pakkasadu-Mulakkadu in Tamilnadu (Sarkar et al., 1995a, b) are potential with respect to LREE (La, Ce). The carbonatite members of Ambadongar area are represented by sovite, ankeritic carbonatite, sideritic carbonatite and silico-barito-fluro carbonatite. The major mineral phases in these assemblages are calcite, siderite, ankerite, fluorite and barite, while the minor phases include dolomite, rhodochrosite, phlogopite, ilmenite, magnetite, pyrochlore, monazite, bastnaesite, anhydrite and strengite. Chief minerals of Samalpatti and Pakkanadu-Mulakkadu carbonatite complexes are calcite, dolomite, allanite and the minor phases comprise dolomite, siderite, ankerite, perovskite, pyrochlore, monazite, bastnaesite, barite, sphene phlogopite, and epidote (Krishnamurthy, 1988).

The most important and well known pegmatite belts in India rich in beryl and lepidolite are located in Rajasthan, Bihar and Andhra Pradesh (vide Fig. 6.1). Comparatively new pegmatite belts of lesser magnitude, but hosting important rare metals—Nb, Ta, Li, Ba, Cs—with or without tin occur in Madhya Pradesh, Orissa, Karnataka and Tamil Nadu. However, tin bearing pegmatites of southern Bastar contain predominantly amblygonite with some lepidolite. Significant quantites of spodumene were reported by Banerjee et al. (1987, 1994) from the Marlagatta and Allapatnary, Mandya district, Karnataka which has also produced significant quantities of niobian-tantalite and beryl. In recent years, Nb-Ta bearing pegmatites have also been discovered in Gujarat.

The rare metal pegmatites are mainly products of highly differentiated fertile granite, which occur very often outside the pluton and are rarely seen within the main granite body. Most commonly the host rocks for these types belong to the low-pressure amphibolites to upper green schist facies which formed at 500-650 °C temperature and at 2-4 kb pressure. Pegmatite derived from fertile granites with high silica, low calcium, variable K_2O/Na_2O, higher Rb, low Sr with high field strength elements like Cs, Nb, Ta, Zr, REE, Y have been recognized as potential source rocks for rare metal mineralization (Ramesh Babu et al., 1993). The spatial and temporal association of such granitic rocks from Precambrian countries of the world needs to be identified in new terrains that are being taken up for detailed study. Anomalous concentrations of certain elements and elemental ratios such as low K/Rb in feldspar, and low K/Rb, K/Cs and Mg/Li in muscovites have also been helpful in locating rare metal pegmatites.

Recent work in Russia and Canada revealed that analysis of certain elements in minerals like Co in biotites (upto 4%) and Rb in feldspar (upto 2.62%) has been recognized as a source for these rare metals (Teertstra et al., 1998). In India, rare metal pegmatites have been the mainstay for columbite-tantalite, beryl, lepidolite and spodumene. Of all the carbonatites known so far, pyrochlore is found in significant quantities only in three carbonatites, namely, Sevakur, Tamilnadu, Sung valley in Meghalaya and Saurchangi in Assam.

The principal REE mineral in Singhbhum copper-uranium belt is xenotime and although it occurs in minor proportion at many places along the 200 km long early-middle Proterozoic shear zone, it has been found to be concentrated only at the Kannyaluka-Khadandungri area. Apatite-magnetite-xenotime veins occur in the above area, sub-parallel with the foliation in the biotite-chlorite and chlorite-quartz schists, partly albitised (Sarkar, 1970).

6.3 METHODOLOGY

The procedure adopted in this study during the first phase is to prepare a grid of cells of equal size (say, 10 km × 10 km) on tracing paper having 49 cells which is superimposed on the geological map of the known mineralized area (vide Figs 6.2 and 6.3). During this exercise, cells with reported occurrence of economic minerals were marked as control cells and the remaining cells without having any report of mineral occurrence are called "barren cells" as per the available geological map used. The present/absent evaluation of the geological variables for each cell including the control cells (having some reported mineralization) is made so that each cell is matched consecutively with the corresponding variables of all the cells (Talapatra, 2004, 2014). The method, thus, makes a comparative study between the control cells and other cells whose ore potential is to be predicted. Accordingly a unique matrix of matching coefficients was programmatically generated (vide Talapatra et al., 1991) as per the following formula: $M = (P + N)/(P + N + U)$, where M = matching coefficient, P = nos. of positively matched, N = nos. of negatively matched and U = nos. of unmatched cells. Cluster analysis was carried out with this matrix showing the results by means of dendrograms, as shown here (vide Fig. 6.4).

Subsequently, representative samples of fresh bed rocks of the study area in parts of Chotanagpur Gneissic Complex (CGC) on the northern parts of Purulia district,

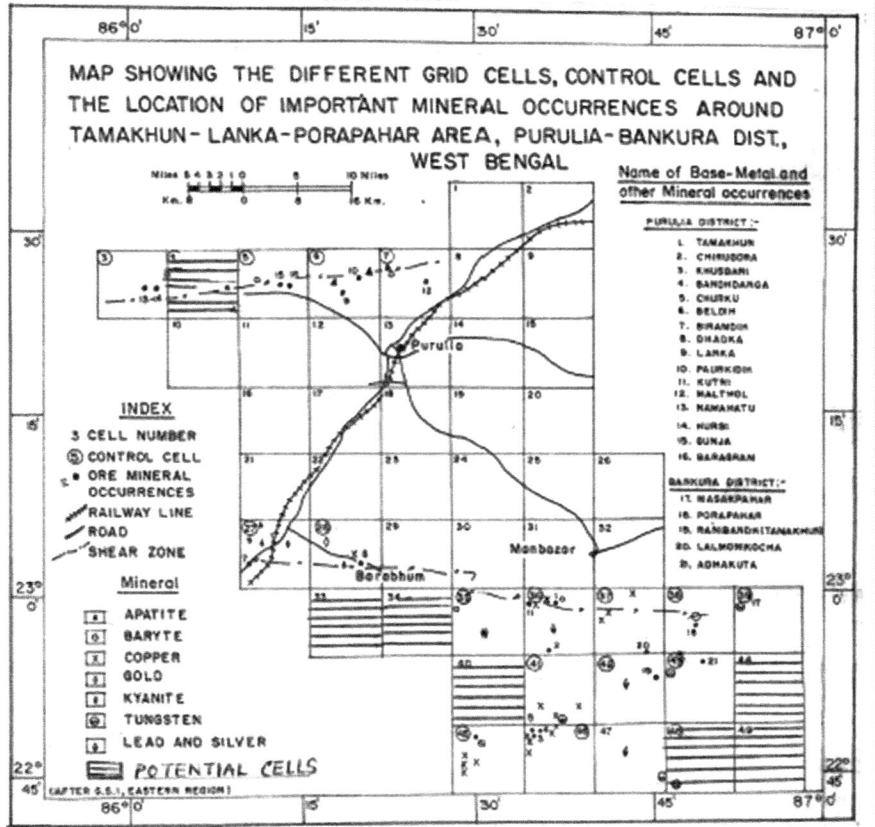

Fig. 6.2 Map showing the different grid cells, control cells and the location of important mineral occurrences around Tamakhun-Lanka-Porapahar area, Purulia-Bankura Dist., West Bengal

West Bengal were collected for detailed geochemical analysis. Granitic rocks of the study area occupy about 80% of the total area within the portion of the CGC falling in West Bengal. The composite granite gneiss with enclaves of metasedimentaries traversed by innumerable small bodies mostly of E – W trends, has been noticed in the southern part of the study area. A medium grained pink gneissic rock has been found to occur along the E-W trending narrow dislocation zone passing from north-west of Kotshila to Sindri and beyond (Fig. 6.3), which marks a distinct shear zone (called North Purulia Shear Zone). Within this gneissic rock, a number of small fertile pegmatite bodies with rare earth bearing minerals along with barite, epidote, magnetite etc. have been located. At the northwestern extremity of the study area, metasediments (sillimanite/garnet bearing mica schists, amphibolites, calc-silicate rocks, etc.) were found to be co-folded along the medium-grained grey granite gneiss (cf. Mazumdar, 1988; Ghose, 1992). These folded metasediments form

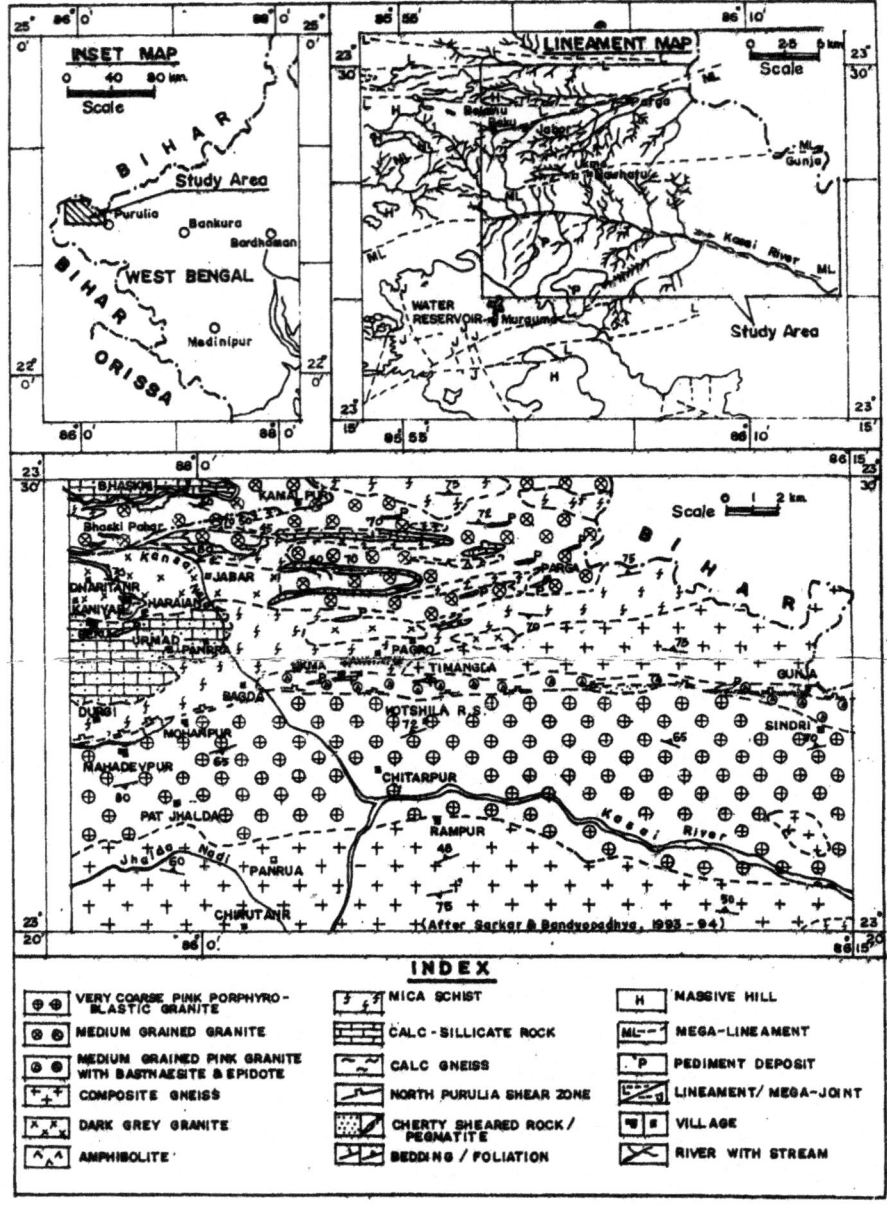

Fig. 6.3 Geological map of Chotanagpur Gneissic Complex around Timangda, Purulia district, West Bengal with a location map of the study area showing the North Purulia Shear Zone and Tamar-Porapahar Shear Zone and a Lineament Map of the region interpreted from the Landsat Imagery

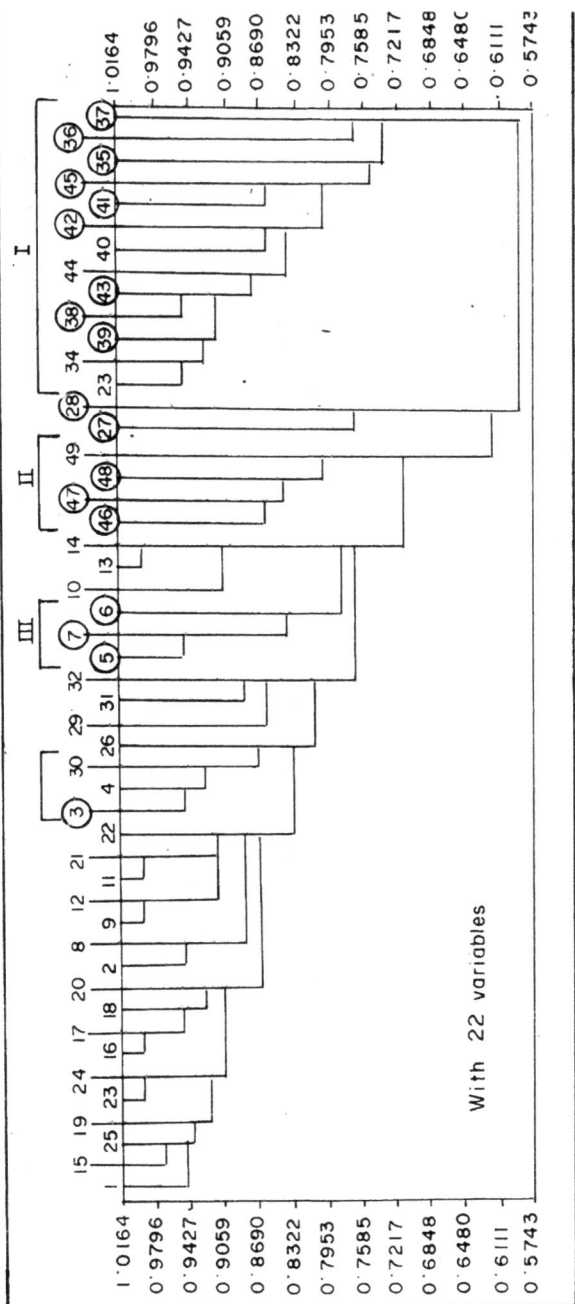

Fig. 6.4 Dendrogram linking the cells of different similarity levels using 22 variables of Purulia-Bankura Shear Zone area

distinct ENE-WSW trending strike ridges. Numerous pegmatite bodies have been found to be emplaced in this area, which are generally parallel to the ENE-WSW trending shear zone. Some of these are fertile in nature and were sampled for chemical analysis.

6.4   GEOLOGY OF PURULIA-BANKURA SHEAR ZONE AREA

The district of Purulia in West Bengal is located on the eastern slopes of Chotanagpur plateau; Bankura district lies to the east of Purulia. A number of base metal and other mineral occurrences reported from Purulia-Bankura area are along the shear zones. Tamakhun is one of the prospects that is thoroughly explored. Similarly, Chirugora-Kutni-Beldi is another such zone where base metal occurrences are reported. Besides, there are also some minor occurrences of base metals in parts of western Midnapur and Bankura districts. Hence Purulia-Bankura shear zone area of West Bengal may be considered as an important belt of base metal and other mineral occurrences marked by the presence of a few clusters of such occurrences of Cu, Pb, Sn, apatite, barite, tungsten, REE etc. (Fig. 6.3).

The area under review essentially comprises Precambrian metamorphites, except a limited area to the north-east where sedimentary rocks of Gondwana age predominate. Unconsolidated Quaternary sediments of Recent to Sub-recent age are of restricted occurrence and are mostly confined to the river channels of narrow width.

The metamorphites are represented by Chotanagpur granite gneiss, biotite granite gneiss, calc-granulite, ultrabasics, metabasics, meta-sedimentaries including crystalline limestone, garnetiferous sillimanite biotite schist (kyanite and graphite bearing at places), pegmatite and quartz vein. The granites have been emplaced along the cores of folded hornblende-schists and schistose quartzites; the pegmatites have been mostly along low dipping limbs of such folds on the hanging wall of the granite massifs. About 90% of the known rare metal and REE minerals bearing pegmatites of the area are confined to this zone. Favourable zones containing these minerals in the pegmatites have been demarcated along with columbite-tantalite. Mukhopadhyay (1989) discussed the various tectonic and genetic problems of the Chotanagpur Gneissic Complex which extends along the study area by introducing plate tectonic model. The Gondwana rocks occurring in the north-eastern part of Purulia district comprise a thick pile of sandstone and shale with coal seam.

The general E-W strike with moderate to high northerly dips of the formations is characteristic. The regional structure consists of close isoclinal folds, in which the fold axes are either horizontal or plunge at low angle towards east or west. Along a narrow zone in north and north-western part of the area the meta-sediments generally dip at moderate angles southwards, and at places dip northwards. The south–dipping narrow zone is a limb of a knee fold, the north limb of which dips below the calc-silicate rocks and amphibolite.

Two sub-parallel shear zones trending roughly E-W occur in this area, which themselves are sub-parallel to the Singhbhum shear zone lying further south. The southern shear zone of the area lies about 50 km towards north of the Singbhum shear zone. The northern shear zone, about 50 km long stretches a little north to Belma in the east and extends as far west as little south of Jhalda. The southern shear zone about 46 km long, starts from about 8 km south of Balarampur, and runs through Beldih, Barabhum etc. to a little south of Manbazar and further east. Evidences of mylonitisation, base metal mineralization and apatite mineralization (including a major occurrence near Beldih) have been noted within these two shear zones (Anon, ibid).

6.4.1   Lineament Configuration

The terrain under study exhibits undulatory topography with a gentle gradient towards south. At the southern parts, the minimum elevation noted is about 240 m above mean sea level (msl) near Kasai River, whereas towards north the maximum elevation of about 320 m above msl was noticed.

The Parga Nala trending towards ENE is the main tributary of northern part, which flows roughly along a major lineament that has been picked nicely from landsat imageries. This lineament is well marked on surface by the strike ridges trending ESE-WSW and having characteristic features of shear zone rocks. In general, a distinct trellis pattern of these nalas and associated rivulets may be discernable at the northern part (vide inset interpreted lineament map in Fig. 6.3). Intermittent bad land type topography was also noticed along this part.

Towards south, the area is drained by roughly E-W trending Kasai River which appears to follow another mega-lineament confirmed by imagery study. The Kasai, along with its tributaries, is one of the principal drainage system of the Purulia district, flowing from west to east which also shows trellis type drainage. The lineament along Kasai is, however, out of the perview of present study. In between the above two mega-lineaments, another E-W mega-lineament has been located through the high ground which is acting as a water divide. This coincides with the trend of North Purulia Shear Zone (NPSZ) showing characteristic features of shearing, silicification and mineralization. Some minor linear features, generally sub-parallel to the above mentioned mega-lineaments, are also observed in the lineament map (Fig. 6.3) of the study area within the CGC which generally do not show any mineralization.

The mega-lineaments described from the study area appeared to have been resulted from deep-seated crustal/tectonic activities which have cut across rocks of diverse composition and structure accompanied by the emplacement of younger intrusive and related mineralization. Ground follow up studies along these areas further indicated these features.

Table 6.1:   List of variables: Purulia-Bankura area

Sl. No.	Description	Sl. No.	Description
1.	Alluvium/soil	12.	Phyllite, mica schist, sillimanite-biotite schist
2.	Gondwana Formation		Quartzite
3.	Vein quartzite	13.	Shear zone
4.	Pink granite	14.	Fault
5.	Porphyroblastic granite	15.	Apatite ore
6.	Granite gneiss, biotite gneiss	16.	Baryte ore
7.	Amphibolite and metabasic rocks	17.	Copper ore
	Dalma lava	18.	Gold ore
8.	Utrabasic rocks	19.	Lead and silver ore
9.	Anorthosite	20.	Tungsten ore
10.	Calcgranulite	21.	Kyanite ore
11.		22.	

6.4.2   Cluster Characteristics

Same procedure was applied for collecting the present/absent data (vide Talapatra et al., 1991) using different variables on the geological map of part of the Survey of India degree sheets 73I and 73J, covering the area under study which includes 49 such cells (Fig. 6.2). Twenty two geological variables, listed in Table 6.1, have been used for this exercise and the present/absent data are tabulated. Same process of matching co-efficient matrix generation has been repeated. Then it is subjected to cluster analysis and finally, a dendrogram linking the cells of different similarity levels is obtained using the computer programme taking all the 22 variables at a time (Fig. 6.4).

This diagram shows that the control cell numbers 5, 6 and 7 through which the Northern Shear Zone of Purulia runs, has been clusterd together showing high similarity level. Control cell no. 3, however, has not been included in this cluster and forms a cluster with cell no. 4. Another good cluster which links 13 cells of Purulia-Bankura area (nine of which are control cells) includes the Southern Shear Zone and also some area lying to the south of it. In this cluster, nine control cells, namely 35 to 39, then 41, 42, 43, and 45 are clubbed together (vide Fig. 6.4) in which important mineral deposits like Tamakhun, Kutni, Masakpura, Porapahar etc. are situated. The remaining four barren cells, namely 33, 34, 40 and 44 appear to have some potentiality from the point of view of locating economic mineral deposit. This large cluster of cells shows some smaller sub-clusters like 39, 38 and 43, which include the control cells of Bankura area. Control cell nos. 27 and 28, and 46 and 47 also form good cluster with high similarity level. Association of cell nos. 48 and 49 with this cluster mentioned earlier, shows some possibility of finding mineral occurrences in these two cells.

The above study has successfully identified the clusters of cells having known mineralized occurrences falling in parts of Purulia and Bankura areas in spite of the limitation of selection of the variables, and has brought out a few apparently "barren cells" clustered with the cells which are part of known mineral occurrences, indicating the possibility of some mineralization within these barren cells. There is every possibility of finding concealed mineralization in such barren cells, provided geochemical exploration with modern technology is conducted in such terrains after taking suitable samples.

6.5   NATURE OF RARE EARTH AND RARE METAL MINERALIZATION

Representative samples of pegmatites from the study area described above have been analysed for the trace elements including Li, Rb, Cs, Ba, Sr and REE, which have been reported by Talapatra et al. (1995). It appears from the results that these pegmatites can be sub-divided broadly into: (i) pegmatites barren of rare metal mineralization, and (ii) fertile pegmatites containing rare metals like Cs, Rb, Li and Rare Earths.

Most of the pegmatites of the area are of the first type. These coarse, pink feldspar bearing pegmatites are frequently occurring in the study area within the two main variants of CGC namely, composite gneiss and very coarse, pink porphyroblastic granite gneiss. Besides pink-feldspar, these pegmatites contain quartz, plagioclase, muscovite, almadine garnet and minor magnetite (Sarkar and Bandyopadhyay, 1994). The second type i.e., fertile pegmatites of the area are complex in nature showing zoning. Because of the significant rare metal and rare element (mainly LREE) concentrations in these pegmatite bodies, these have been shown in the geological map (Fig. 6.3) with slight exaggeration. This may be again classified into two distinct types from their geochemical signatures (vide Cerny, 1991). These are: (i) Li, Rb, Cs, Ba, Sn, Nb, Ta, etc. bearing pegmatites (LCT type), and (ii) REE, Nb, Y, Th, Ti, Ba, F, etc. bearing pegmatites (NYF type). The LCT type is largely associated with paraluminous granites, whereas the NYF bearing latter type of pegmatite is generally genetically related to alkaline granites with carbonatitic affinity (Talapatra et al., 1995). Cs the most active metal on earth is present in LCT type pegmatites with pollucite, rhodizite and lepidolite found in association with rubidium.

It has been observed that the REE bearing pegmatites, occurring along the shear zone are associated with medium-grained, gneissic rock with epidote veins. A few small carbonatitic bodies have been located close to these pegmatites by subsequent investigations. The presence of rare-earth bearing mineralization in the form of allanite-bastnaesite along with coarse barite was observed in an abandoned quarry near Nawhatu (Sarkar et al., 1995b; Sarkar and Bandyopadhyay, 1994). The brownish black heavy lumpy material contained mostly allanite, (?) monazite, with bastnaesite veins. Beside these minerals, opaque oxides (mostly rutile and magnetite) are also present. Near Timangda, the epidote-bearing, pink gneissic rock along the E-W trending NPSZ also contains similar pegmatitic rocks having barite-allanite-bastnaesite-epidote-magnetite veins. These are 5 to 10 cm wide and can be traced for a distance of 10 to 12 m. The rare earth mineralization is also noticed within the pegmatitic rocks near Gunja, where allanite and bastnaesite are present in association with barite. It is apparent that barites are closely associated with rare earth bearing NYF type pegmatite along this lineament of Purulia area. Analytical results of nine pegmatite samples containing allanite-bastnaesite association showed REE varying from 0.5 to 35.23% as reported by Talapatra et al. (1995).

The LCT type pegmatites of the study area essentially contain quartz, plagioclase, mica (both colourless and pinkish), pollucite and spodumene. Chemical analysis results of 13 such pegmatite samples of Beku and Jabor area have shown significantly high values of Cs (varying from 50 ppm to 13.73%), with moderately high Rb and Li (Talapatra et al., ibid). Pollucite is characteristically found to be present within pegmatites having pink variety of lithium bearing mica. It is observed that the pegmatitic rocks of ENE-WSW shear zone extending upto Parga is worth sampling for Cs concentration. Both the shear zones described here falls within the control cells nos. 5, 6 and 7 and also in the so-called barren cell no. 4 mentioned earlier.

6.6   CONCLUSION

Application of the deposit modelling technique along the mineralized areas of West Bengal mentioned here has successfully identified the clusters of control cells having base metal and other mineralization, and has brought out a few apparently "barren cells" clustered with these control cells, as mentioned earlier, indicating the possibility of getting some concealed mineralization within these cells also. Despite limitations of choosing significant geological variables due to lack of large scale recent geological maps along the areas under review, application of this method in these mineralized belts provides an excellent picture of the ore potentiality of the area indicating some statistically potential ore mineralized cells (now considered to be barren) for further detailed exploration using different conventional and non-conventional exploration techniques (Talapatra and Bose, 1979; Talapatra et al., 1981, 1986b). Use of this deposit modelling technique utilizing the available geological variables in conjunction with geochemical and geophysical variables after designing suitable grid cells with proper orientation for any mineral reach terrain is highly recommended for quick mineral appraisal of REE and Rare Metal occurrences also on regional scale without much investment along the Indian Subcontinent as shown in Fig. 6.1. This figure shows important REE and RM occurrences within the continental parts of India reported from its different provinces which require immediate detailed study by the exploration geologists for future exploitation of these elements. Even an application of such techniques along the Precambrian terrains of the different parts of the world, especially in southern and southeastern parts of Asia, will be very much beneficial.

Petrochemistry of the rocks of Chotanagpur Gneissic Complex exposed in Purulia-Bankura area have indicated that the crustal evolution of this part of Precambrian shield has experienced some tectonic adjustment resulting in the formation of some mega lineaments with some linear features, generally sub-parallel to this. A number of lensoid pegmatite bodies have been located along or close to these mega-lineaments and emplaced within the gneissic terrain with their longer axis parallel to these linear structures.

Laboratory studies indicated that most of the pegmatites of the area are barren of rare metals and REE, possibly indicating that these are less evolved. However, there are also some highly evolved, complex type of fertile pegmatites which were subsequently emplaced along the mega-lineaments. Among these pegmatites, some are enriched in rare metals e.g. Li, Rb, Cs, etc. which are present along ENE-WSW trending lineament and others are enriched in REE, Y and Nb emplaced along E-W trending lineament following the NPSZ, as already discussed. Presence of REE bearing mineralization within the pegmatites lying between Ukma and Gunja in association with barite definitely suggest that the E-W trending lineament has brought the rare earths from lower tectonic levels and the mineralization is extending here for about 20 km. Mineralization of this type is expected to be present in other parts of this mega-lineament also. Similarly, rare-metal bearing fertile pegmatite bodies, noted along the ENE-WSW trending mega-lineament joining approximately Beku and Parga, suggest the likely presence of similar pegmatite bodies along the extension areas of the lineament. Presence of carbonatite bodies along the shear zone mentioned earlier also indicates such possibilities.

The area under examination appears to be an ideal site of lineament-controlled metallogenesis as a result of the Precambrian crustal evolution in this part of the Indian shield. Detailed study of these fertile pegmatites in association with the carbonatite bodies along the prominent lineaments of the area is expected to unearth sizeable concealed deposits of rare-earths and rare metals which are very useful as raw materials in high tech products necessary for modern living and also for other important mineral-based industries. Application of identical approaches of mineral exploration in other Precambrian mineralized belts with pegmatites of India (vide Fig. 6.1) and other similar Precambrian terrains of the world which are associated with pegmatite-carbonatite assemblage is likely to be highly rewarding, specially along with marked lineaments passing through prominent shear zones.

Chapter 7
STANDARD ANALYTICAL METHODS FOR GEO-SAMPLES

7.1 INTRODUCTION

Geological samples may be of different types, namely fresh bed rock, weathered/gossanised rock, *in situ* transported soil, stream sediment, plant etc, as already mentioned earlier. Depending upon the nature of the sample collected, the technique of chemical analysis is selected. The different types of geochemical samples collected as per specifications during the field survey and exploration programmes are analysed in the various chemical laboratories. The quantitative analytical data of individual elements so generated may be fruitfully used for preparation of different types of geochemical contour maps for future use in connection with geochemical prospecting and other geoscientific and environmental purposes. Such data are generally used to prepare geochemical atlas for the individual elements for the land area studied from where samples are collected. At present this type of geochemical atlas is available for most of the developed countries of the world, where such atlas is used by the town planners, architects, mining engineers, agriculture and other country development engineers. In due course such geochemical atlas will be in demand for most of the developing countries of the world. In the following paragraphs, various types of chemical analyses using different type of sophisticated instruments will be narrated in short. One thing should be clearly noted that depending upon the type of analysis, quantity and nature of sample should be collected after consulting the analytical chemist, who is the best judge for selecting the types of analysis to be followed with samples.

© Capital Publishing Company, New Delhi, India 2020   165
A. K. Talapatra, *Geochemical Exploration and Modelling of Concealed Mineral Deposits*, https://doi.org/10.1007/978-3-030-48756-0_7

7.2   SEMI-QUALITATIVE AND QUANTITATIVE ANALYSIS OF GEO-SAMPLES

Depending upon the nature and mineralogical composition of the geological sample collected, analytical chemist will decide about the type of analysis to be conducted for the sample. Accordingly, while sending the geo-samples to the Laboratory, chemical analyst should be consulted giving him the rough idea about the location of the rock samples and likely mineral assemblage of the sample so that he can also think about the geological problem to be solved. Depending upon the mineralogical composition of geo-material one can select the suitable methods for qualitative and, if required, quantitative analysis. The first two analytical methods described in the following paragraphs are generally used for qualitative data followed by the four different quantitative methods described subsequently. In this respect the Geologists should interact properly with the analyst so that the results can be fruitfully utilised.

7.2.1   Geochemical Analysis by Visual Colorimetry

This analytical technique is suitable for rapid semi-quantitative determinations of Cu, Pb, Zn, Ni, Co, As, W and Mo in soils, rocks, stream sediments etc. and is very useful for checking and quick disposal of large number of samples that are collected in the field. Apparatus needed is simple and with some practice reliable results may be obtained. Overall precision is about 30%. The sample is decomposed by digestion with acid or fusion with potassium pyrosulphate/alkali and extraction is done with dilute acid. Aliquots are taken from this solution to different elements by developing specific colour and comparing it with that developed with calibration solution. 0.1 gm sample (80-100 mesh) is properly decomposed by acid digestion or alkaline/acidic flux fusion and the digested or fused mass is extracted with water or acids. The aqueous/acidic extract is masked by suitable masking agents for interfering radicals and by suitably adjusting pH. Now, a suitable organic chelating agent like dithizone and furiedioxime, etc. is added to the sample solution to have a coloured compound with one of the wanted elements. The coloured species is extracted in a definite value of a suitable organic solvent and the colour intensity of the solvent layer is then visually compared with that developed with calibration solutions. In this endeavour the chemist will help and guide the field geologist.

The lower level of detection (LLD) for this method is 10–50 ppm. However, with suitable chemical preconcentration LLD may be brought down to 1 ppm or less for some elements like As, Mo, W etc. In general this method normally gives semi-quantitative results.

7.2.2   *Geochemical Analysis by Spectrograph*

A spectrograph is a versatile instrument with the help of which samples of diverse nature and widely varying composition can be analysed for simultaneous estimation of broad spectrum of elements with speed, at low cost and with tolerable accuracy. In routine semi-quantitative analysis the precision is low. But still the vast information, which is gathered in short time outweighs the above minus point and it is an important method of analysis in advanced laboratories in geo-analysis. This analytical technique is unparalleled in qualitative detection of elements and hence may be used for quick scanning of samples for geochemical surveys.

   In this technique 10-15 mg (–80 mesh) sample is thoroughly mixed with equal weight of pure graphite powder and the processed sample is arced or completely burnt between two graphite/carbon electrodes, the arc light passed through a dispersion medium and isolated radiation being recorded on photographic plate. Standards are also arced similarly on a different plate. The spectrum of line-rich material like Fe is also arced and recorded simultaneously. The sample plate and standard plate (also containing the Fe spectrum) are placed in juxtaposition on a visual comparator; the position of line compared to spectrum of reference of material like Fe indicates qualitative presence of the elements and amount of blackening of plate emulsion is indicative of concentration of the element present. By comparing the corresponding spectral line of a specific element in the standard sample plate, it is possible to determine the elements in ppm level. The LLD for volatile elements is of the order of 100-150 ppm and for non-volatile elements it is 10 ppm.

7.2.3   *Geochemical Analysis by AAS*

The Atomic Absorption Spectroscopy (AAS) is a spectro-analytical procedure for the quantitative determination of chemical elements using the absorption of optical radiation by free atoms in the gaseous state. Atomic Absorption Spectroscopy is based on absorption of light by free metallic ions. The AAS analytical technique is most suitable for quantitive determination of large number of elements in soils, rocks, sediments etc. and is very useful for rapid checking and quick disposal of large number of samples. In AAS analytical technique, the sample solution is aspirated into an air-acetylene or N_2-acetylene flame for complete atomization of the test materials. Now, absorbance of a characteristic radiation by the atom population of the desired element is measured and from these absorbance value the elemental concentration is computed.

   The accuracy and precision by AAS is high and good. Usually +1%-5% accuracy is obtained for samples in major quantity and 2% to 15% in the minor and trace level. A number of elements (about 30) can be analysed from one solution, one after another. In this technique a definite weight of the sample is decomposed by digestion with a mixture of acids and individual elements are measured by atomic absorption spectrometry using different hollow cathode lamps and proper background correction.

For rapid geochemical screening by AAS, 0.5 gm sample is treated with 5 ml of hydrochloric and nitric acid mixture and digested in an alluminium rack placed over hot plate. The digested mass is allowed to cool and volume made to 20 ml. Now AAS instrument reading for water is set to zero absorbance. The calibration of solutions for the desired element is aspirated into the instrument followed by the test solutions and the blank. The respective absorbance readings are recorded and from these absorbance values the elemental concentration in the test solutions are computed by an analyst.

7.2.4 Geochemical Analysis by X-ray Fluorescence (XRF)

The X-ray Fluorescence (XRF) is a non-destructive analytical technique used to determine the elemental composition of materials. XRF analysers determine the chemistry of a sample by measuring the fluorescent x-ray emitted from a sample when it is excited by a primary X-ray source. XRF analytical technique makes the use of high energy X-ray photons from an X-ray generator to excite secondary (fluoroscence) characteristic x-rays from a geo-specimen. The characteristic line spectra emitted by different elements contained in the geo-specimen are then differentiated by crystal grating and finally detected by suitable detectors. The intensity of each line is proportional to the concentration of each individual elements. The overall precision is very high and the technique is now widely used for routine analytical work.

The precision in this type of analysis is 10% for trace elements and 2-3% for major elements. In the field XRF unit, a mains/battery operated portable analyser instrument based on isotope technology may be used. The measurement is based on energy dispersive X-Ray fluorescence. The radiation from a radioactive source strikes the sample and excites fluorescent radiation. The spectrum of fluorescent radiation i.e. the intensity versus the energy of the radiation contains the information required for the analysis of the sample. The result is processed by a microcomputer in the electronic unit. In quantitative XRF analysis, however, much pretreatment of sample is carried out before measurement of X-ray intensity of the desired element.

7.2.5 Geochemical Analysis by ICP-AES

Inductively Coupled Plasma Atomic Emission Spectroscopic analysis (ICP-AES) also referred to as inductively coupled plasma optical emission spectrometry is used for the detection of chemical elements. This is one of the optical emission methods. Here the excitation source is an inductively coupled plasma. The plasma used is an argon gas. Induced field from a radiofrequency current through a coil outside three concentric quartz tube, causes the argon-gas flowing inside, to be excited enough to produce a high temperature zone varying from 6000 to 10000 °C. Analyte either in solution (aqueous or non-aqueous) or finely powdered is aspirated as aerosol and passed through the high temperature zone of the plasma. Atoms in the analyte get excited and emit characteristic radiations. After resolving the emitted radiation spectra by a suitable dispersing medium (grating), intensity of the characteristic line of any element can be measured by a photo-multiplier tube and matched against standard to determine the concentration of the element in the sample. Many wave lengths of varied sensitivity are available for the determination of any one element, so that the ICP is suitable for all concentrations from ultra trace level to major components. The overall precision is very high. About 40 elements can be analysed by this technique at a time. REE can also be estimated by ICP-AES. By this technique As, Sb, Se and Te can be determined directly in the sample solution or after their enrichment as hydrides. Platinum group of elements can also be determined by fire Assay cum ICP-AES and semi-quantitatively by spectrograph (OES) after chemical enrichment.

7.2.6 *Instrumental Neutron Activation Analysis*

The technique of Instrumental Neutron Activation Analysis (INAA) is based on utilizing high energy neutrons for irradiation of a sample to give off gama radiation which is then analysed for detection of elements. This is a nuclear process used for determining the concentrations of elements in a vast amount of materials. INAA allows discrete sampling of elements as it disregards the chemical form of a sample and focusses solely on its nucleus. The technique of neutron activation analysis is based on measurement of radiation released by the decay of radiometric nucleus formed by neutron irradiation of the material. The samples that can be analysed with this method stem from a number of different fields including biology, geology, chemistry, environmental science etc. The method is especially suitable for trace elements and REE analysis where accuracy and precision are of importance (Gordon et al., 1968).

7.3 CHOICE OF CHEMICAL ANALYSIS

The choice of analytical procedure for geochemical samples is quite important because of the limitations of some sample digestion procedures and the characteristics of different analytical instruments (Fletcher, 1981). It is generally desirable to analyse for 'total' rather than partial metal contents, since active chemical dispersion processes have seized in most of the Precambrian terrain and the dispersed metals tend to be held within the lattices of fairly resistant minerals like hematite and goethite rather than being absorbed on clays and amorphous iron oxides. However, continued experimentation and orientation surveys are essential before taking any decision. In this regard choice of analytical technique may be done by searching the suitable analytical methods on line for the different types of samples from the "Google", which gives the details of different analytical procedures that are available now-a-days. In fact techniques of analytical methods are improving day by day.

With the introduction of ICP-methods, atomic absorption analysis will be directed more towards specific elements best done by that technique. X-ray fluorescence (XRF) remains almost unrivalled for major element analysis of fresh rocks (Levinson and de Pablo, 1975). But ICP analysis provides an alternative at a price that makes major element analysis of exploration samples feasible. In addition some elements like U, Th, Bi and Sn are most suited to XRF analysis. However, it may be mentioned here that analysis of geochemical samples basically requires an anomalous value along a sample line so that relatively high value of the element of analysis is important, rather than its absolute value. Among the different types of chemical analysis mentioned and described here in short, INAA appears to be quite suitable for determination of REE and different trace and major elements present in geochemical samples for detection of concealed deposits of economic minerals within terrestrial and offshore mineral deposits of the world. As such choice of analytical methods will depend upon the type of sample to be analysed, its major and trace-element contents and its anomalous presence in the sample area. Depending upon all these factors, suitable choice of analytical technique will not be a difficult task taking appropriate suggestion from analytical chemists.

It may be mentioned here that some sophisticated instruments are now-a-days available for analyzing different types of samples, namely, Scanning Electron Microscopes (SEM) which produce three-dimensional (3D) images, while Transmission Electron Microscopes (TEM) only produce flat (2D) images. These types of electron microscope have emerged as a powerful tool for the characterization of a wide range of sample materials. Their versatility and extremely high spatial resolution render them a very valuable tool for many applications. TEM has much higher resolution than SEM. Out of these two types of electron microscope, SEM is based on scattered electrons, while TEM is based on transmitted electrons. TEM give informations about the internal structures, while SEM is better for 3D surface morphology. Another modern instrument is SIMS which means Secondary Ion Mass Spectrometry, which is a technique used to analyse composition of solid surfaces and thin films by sputtering the surface of the specimen with focused primary ion beam. The sample surface is sputtered/etched with a beam of primary ions (usually O_2^+ or C_S^+) for high precision bulk analysis.

Chapter 8
SUGGESTIONS FOR FUTURE WORK

8.1 INTRODUCTION

Major parts of the terrestrial areas of the earth are covered by soil, alluvium with very little *in situ* rock exposures. As such possibilities of the presence of concealed mineral deposits are quite high in such covered areas. In the chapters described so far, suggestions for geochemical exploration of concealed terrestrial and offshore mineral deposits have been outlined with examples. In this effort mineral deposit modelling has been successfully applied here. The aim of mineral deposit modelling is to build up the reality of the exploration of concealed deposit, if any, as closely as possible using all the available surface and subsurface data. It replicates the body geometry of the concealed mineral deposit along with the physical distribution of various spatial parameters in order to assess the 3-D configuration of the mineral deposit. Both estimation and simulations enable us to realize the extent of the deposit as well as many such aspects of reality.

Mathematical techniques of modelling of mineral deposits include both (i) deterministic and (ii) probabilistic approach (Sarkar, 2005). Each of these techniques has its own advantages and limitations. The techniques under deterministic group include (i) determination of average of all samples within a geological domain, (ii) linear interpretation (iii) inverse distance weighting, (iv) contouring, and (v) moving average and trend surface analysis (Krumbein and Graybill, 1965; Koch Jr. and Link, 1970; David, 1980 and Jhonson, 1995). The techniques under probabilistic group include some of the recent developments like (i) classical statistics (Sarkar et al., 1995a, b), (ii) geostatistics, (iii) spectral analysis, (iv) simulation and application of Zipf's law (Zipf, 1949), (v) baysian statistics, (vi) logistic regression, (vii) belief function, (viii) decision support system etc. Narration of all the above techniques is beyond the scope of this write-up. Number of reference books are

© Capital Publishing Company, New Delhi, India 2020   171
A. K. Talapatra, *Geochemical Exploration and Modelling of Concealed Mineral Deposits*, https://doi.org/10.1007/978-3-030-48756-0_8

available. However, with the fast developments of information technology (I.T.), it is now possible to analyse huge number of the data with repetitive calculations to arrive at a statistically significant conclusion, which are available in some text books and research papers, mentioned earlier.

8.2   SUGGESTIONS FOR INDIAN EARTH SCIENTISTS

It has been emphasized that a large part of the Indian sub-continent is covered by soil, sand, alluvium etc. in the land areas. In fact, continuous exposures of fresh *in situ* rocks are very difficult to locate in such areas. This is true for most of the countries of the world where concealed mineral deposits of various kinds are likely to be present. However, attempts have been made here to suggest some techniques, so that the concealed deposits in land areas and/or offshore areas within the territorial waters of EEZ may be located either along strike continuity of known deposits or along some new tracts where, some significant field data/parameters help to delineate the target areas by applying some statistical methods with qualitative/quantitative data or parameters, provided sufficient data are available. It has been observed that different types of geological, geophysical and geochemical data are being successfully utilized to locate concealed deposits in parts of land areas. Some Indian examples of base metal etc. have been cited from Precambrian land areas as well as along some offshore heavy mineral placers formed during Quaternary or Recent times along coastal areas, especially with concentration of REE and RM content.

In this regard, G.S.I. which is one of the oldest survey of the world, is actively engaged in the exploration of different mineral deposits of the country. Very recently G.S.I. is also using ultra modern remote sensing technology with the help of National Aeronautics and Space Administration (NASA), U.S.A. This sensor is an engineering marvel, which has been proved effective for mapping surface mineralogy in different parts of the world. G.S.I. is going to utilize these air-borne hyperspectral data by their geologists to find surface signatures of mineralization in fourteen promising mineralized blocks of India within next three years. The Atomic Minerals Directorate (AMD), Govt. of India is also running one of the country's leading multidisciplined and multifaceted exploratory organization, especially for uranium, thorium, rare metals (Nb-Ta, Be, Li etc.), rare earths (La, Lu etc.) and yttrium. It has been playing a pivotal front-end role in the country's atomic energy and related programmes through exploration for deposits of atomic minerals etc. required for nuclear and other hi-tech industries like electronics, telecommunications, information technology, space, defence etc. for more than fifty years. The targets for exploration for RM and REE resources are essentially the primary pegmatite source as well as coastal placers and elluvial concentrations derived from these, which are being worked by both GSI and AMD.

In the land areas, search for mineral deposits is generally taken up after systematic geological mapping of any terrain, provided there is some significant indication of ore mineral occurrence/concentration in the area. This is followed by large scale

mapping (1: 25.000 scale) of the area of interest after making photogeological studies in conjunction with other available remote sensing data. If there is any positive
indication of mineralisation, systematic geochemical sampling on large scale map,
along with ground geophysical studies is taken up to locate the anomalous zones
which may indicate the tentative configuration and extent of the causative ore mineral body along certain belt. In such condition the geochemical atlas of the area
under investigation for the individual elements like Cu, Pb, Zn, etc. are quite helpful
for locating the anomalous zones with high contour values. Subsequent subsurface
probing of these zones may establish a mineral deposit. For a virgin area, prior study
of remote sensing imagery and air-borne geophysical data may help in locating the
area of interest before conducting systematic geological investigations with proper
sampling. This is followed by ground geophysical survey and geochemical sampling. Mineral investigation in a deposit for locating its subsurface configuration are
generally conducted through exploration methods like pitting, trenching, drilling,
exploratory mining etc. Once a mineral deposit is located in an area, *'mineral
deposit modelling'* may be taken up for the entire region or belt for resource evaluation along the area as detailed earlier. This may be systematically applied in any
part of crustal areas of the world for finding concealed mineral deposits.

   It has been observed that utilizing different parameters, deposit modelling may
be possible for predicting the continuity of the hidden ore body, if any, along the
land areas. This is also possible for the heavy mineral placer deposit etc. in the offshore areas of EEZ with the help of different available softwares taking into consideration the vibrocore samples of different blocks collected from different cruises
conducted along the coast as reported earlier. Sometime this type of computer-based
studies may even indicate new target areas from virgin areas. Computer-based
3D-diagrams from offshore areas upto the EEZ of any country from vibrocore samples give sufficient indication of concealed deposits of the study areas as well as its
surroundings, as discussed in details in Chapters 1 and 2 with diagrams.

   Mineral exploration in terrestrial deposits generally requires different concepts
depending upon the genesis and control of emplacement of the ore body, which may
be stratabound, stratiform or structurally controlled. Depending upon the terrain
condition and geomorphology of the area under study different sampling media are
chosen to decipher the varied nature and types of parameters within the study area,
which may be useful in regional geochemical survey. Stream sediment sampling
technique has been considered as the most suitable media of regional geochemical
survey, especially for generating regional geochemical map depicting essentially
the surficial element distribution pattern (Webb et al., 1978; Plant et al., 1988).
Regional geochemical maps so produced will delineate anomalous zones, which
may be subsequently studied in detail for search of mineral deposits, if any. A series
of geochemical surveys conducted in different parts of the earth during the 60's
showed conclusively that widely-spaced stream sediment samples (at densities in
the range of one sample per 2.5 to 180 sq. km.) could be used not only to provide
fundamental geochemical information relevant to the regional geology, but also to
broadly delineate potential mineralized areas (Webb et al., 1964; Nichol et al., 1966;
Garret and Nichols, 1967).

It has been observed that the main features influencing the geochemical pattern on the maps prepared from such data are the bedrock geology, nature of overburden, contamination and other environmental conditions. In all such studies with fine-grained fraction of active sediments, the elemental composition of the sample was found to be broadly stable under a wide range of seasonal variations and topography. All these techniques of mineral exploration may be successfully applied to Precambrian and other terrains of India and especially of China, South and Southeast Asian countries, in particular.

It may be mentioned here that to start with the geochemical study of any land area, reconnaissance survey with multi-element geochemical analysis is essential. This should be followed by detailed geochemical analysis, provided there are some significant results at initial stage, so that regional geochemical maps of the area can be generated. Special emphasis should be given for exploration of REE and RM which has become an essential commodity as detailed in Chapter 6. Once the regional scale multi-element reconnaissance surveys (Hood, 1979) are over, detailed geochemical studies may be taken up during the second phase of the programme on large scale, say 1: 25,000 scale or more in selected areas based on the elemental distribution pattern revealed by the regional geochemical maps. Such detailed investigations are generally carried out in the follow-up stages of an exploration programme involving bedrock, weather cap rock (gossan), soil, stream sediment, water and plant surveys as is applicable in the local physiographical and geological conditions especially in the case of base-metal sulphides. Similar type of detailed geochemical work may, however, be simultaneously taken up along the extension areas of known mineralized belts/occurrences in any country.

8.3 SUGGESTIONS FOR EXPLORERS OF ASIA AND OTHER SIMILAR COUNTRIES OF THE WORLD

Detailed methodology of geochemical surveys along with different techniques of sample or data collection (applying both conventional and non-conventional methods) has been outlined earlier. Under the conventional techniques, sampling of different media has been discussed and standard methods of sample preparation have been outlined. Study of weathering features have been given due consideration in the context of Indian subcontinent and role of gossan geochemistry in geochemical exploration has been highlighted. Parts of peninsular India, mostly along the western and northern part, are covered by alluvium, wind-borne desertic sand and transported soil where the conventional geochemical techniques are not suitable. As such, a detailed account of non-conventional methods for detection of concealed deposits applying the concept of vapour phase geochemistry, electro-geochemistry and isotope geology has been discussed which will be most suitable for land areas of the world with similar concealed type mineral deposits.

As regards REE and Rare Metal exploration, monazite as accessory mineral is extremely rare in the green schist facies, but common to abundant in the granulite facies rocks (Overstreet and Olson, 1964). That explains how granulite rocks of south and eastern India also act as the source rocks for monazite bearing placers in coastal areas. It is interesting to note that locally the content of monazite in high grade metamorphic rocks can be considerably high. Davidson (1956) reported monazite content of 17.9% in a garnetiferous biotite schist in a migmatitic zone at Tadikarakoram, Kerala.

Warm and humid climates with abundant rainfall are conducive to decomposition of the basement rocks and liberation of chemically stable minerals. High rainfall, accompanied by favourable topography, can transport the resistant and heavy minerals into nearby streams and rivers, which in turn convey these to the sea, mainly as bedload. The Kerala coast is riven by numerous rivers, rivulets and creeks. Annual rainfall in most of the source areas of the Indian placer deposits is about 300 cm or more. Action of waves, wind, tides and current do the necessary winnowing and deposition in such areas. Extreme effect of winds produces dunes as in the Ganjam beach of east coast of India.

Beach placers of India nonetheless are the largest repository of monazite being the same as noted in earlier sections. It is high time now that we make a more serious search for hydrothermal-metasomatic stratabound (or otherwise) bastnaesite + monazite bearing deposits like that at Byan Obo, China, a carbonatitic deposit as at Mountain Pass, USA, or atleast a monazite bearing vein deposit as at Steenkampskrael, South Africa. The possibility of finding a new type of deposit different from any of those mentioned here, cannot be ruled out either. The Indian earth scientists will surely be upto this challenge and this will attract the exploration geologist of any other country where similar type of beach placers are present.

To initiate computer-based geostatistical models in mineral resource estimation requires a structured database described at the beginning in the holistic diagram (vide Fig. 1.1). An exploration database is a repository of different types of data namely, structural and integrated geological, geochemical, geophysical, drill holes and assay information that allows storage and retrieval of data, and controls redundancy to serve one or more computer applications in mineral exploration in an optimal fashion (cf. Talapatra, 2001). A common and controlled approach is used in adding new data, update and validate the existing data for modifying and retrieval of existing data within the database.

The recent trend in the development of computer-based models for mineral exploration includes knowledge-based expert systems, fuzzy systems, decision support system, neural networks and genetic algorithm. Knowledge-based expert systems are primarily based on symbolic processing wherein problem description and their solutions are stored, and then retrieved to solve other similar problems. It has attracted the attention of earth scientists because of recent successful practical application in the field of mineral exploration (David et al., 1987; McCammon, 1986; Sarkar et al., 1988; Dimitri-Kopoulas, 1993).

8.4 CONCLUSION

Before concluding the chapter, let me spell out here that huge amount of different types of concealed mineral deposits are lying below the soil, sediments and other types of cover materials in different parts of the world. Exploration geologists are supposed to unearth these concealed deposits for fruitful use by the mankind using the modern technologies available now, which has been outlined in this book. It may be mentioned here that in the field of mineral exploration, management of huge amount of information resulting from an exploration campaign, which is a multi-stage decision process, point to the need of decision support system. The successive steps of multi-stage decision process for mineral exploration should include (i) reconnaissance geological mapping and collection of geochemical and geophysical data, (ii) regional exploration with large scale mapping and detailed sampling of surface and subsurface data and (iii) exploration of ore body including systematic drilling and exploratory mining. Each of the stages end-up by a decision to continue or not, for further detailed work. The proper investigation of a mineral deposit through computer-based modelling is the foundation upon which the subsequent mine decision (viz. design, planning and economics) are taken. Recent steps of mineral exploration use different techniques like fractal models, artificial intelligence etc. that are not discussed here. However, with the advancement of information technology, its varied applications in the search of concealed mineral deposits may be applied successfully in future for exploration of ore body including exploratory mining etc.

REE enrichment in ion adsorption clays, together with placers, potentially also have the easiest processing routes for extraction of REE. However, environmental issues can be a concern: placer, laterite and ion adsorption clay projects potentially have large foot prints and can have substantial impacts on local environment and communities. Further more, public fear of radioactivity has prevented placer deposits being used as REE resources in some countries.

Placer deposits are also of interest due to the relative ease of processing, as long as the issue of their natural radioactivity can be successfully managed. In fact, it is quite suitable for a country like India where huge area is full of placer deposits along its sea coasts. Accordingly, the countries of the world with such placer deposits along their sea coasts, can definitely try to explore REE and RM deposits. REE production as by-products, such as leaching of REE during bauxite processing or removal of REE during phosphate processing for fertilizer is also of significant interest for future supply of these commodities.

Research into the different types of deposits is still extremely useful because of additional knowledge of REE and RM mineralogy, mobility and concentration that will help define better exploration model. This will definitely point to new exploration targeting suggestions as well as better processing methods, especially for concealed terrestrial and offshore mineral deposits.

Before concluding the geochemical exploration and modelling techniques of concealed terrestrial and offshore deposits, let us discuss the salient mineralized zones of India in short which will also help the explorers of any other similar country of the world. The most important and well known pegmatite belts in India are located in Rajasthan, Bihar and Andhra Pradesh. New Pegmatite belts of lesser magnitude, but hosting important rare metals – Nb, Ta, Li, Be, Cs – with or without tin occur in Madhya Pradesh (M.P), Orissa, Karnataka and Tamil Nadu. Tin bearing pegmatites of southern Bastar, M.P. contain predominantly amblygonite with some lepidolite as reported by Banerjee (1999). Significant quantities of spodumene were reported from Mandya district, Karnataka which has also produced significant quantities of niobian-tantalite and beryl. In recent years, Nb-Ta bearing pegmatites have also been discovered in Gujarat.

Pegmatites of rare element class are mainly products of highly differentiated fertile granite mentioned earlier, which occur very often outside the pluton and are rarely seen within the granite. Most commonly, the host rocks for these types belong to the low-pressure amphibolites to upper greenschiest facies which are formed at 500-650°C temperature and at 2-4 Kb pressure (Banerjee, ibid).

Fertile granites with high silicic, low calcic, variable K_2O/Na_2O values, higher Rb, low Sr and significant enrichment of high field strength elements like Cs, Nb, Ta, Zr, REE and Y have been recognized as potential source rocks for rare metal mineralization. The spatial and temporal association of such granitic rocks from Precambrian terrains needs to be identified in new locales that are being taken up for detailed study. Anomalous concentration of certain elements and elemental ratios such as a low K/Rb in feldspars, and low K/Rb, K/Cs and Mg/Li ratios in muscovites have also been helpful in locating rare metal pegmatites.

It is nowadays recognized that the use of rare metals and rare earths will multiply manifolds because of their utility as a value added product in a number of high technology applications such as nuclear, space, defence, information technology etc. mentioned earlier. It is, therefore, imperative that earth scientists have to be more innovative to discover new deposits of REE and RM, applying several approaches mentioned here for their exploration and exploitation as has been followed in some advanced countries of the world.

In India, Rare Metal pegmatites have been the main source for columbite-tantalite, beryl, lepidolite and spodumene. Carbonatites, for example, constitutes the most important source for Nb in pyrochlore, found in Araxa in Brazil and Leushe in Zaire. Among all the carbonatites so far located in India, pyrochlore is known to occur in significant quantities in three carbonatites reported so far from Sevattur in Tamil Nadu, Sung Valley in Meghalaya, and Samehampi in Assam. Certain pegmatites contain rare minerals associated with cassiterite, as in the case of several rare metal pegmatites of Bastar, Madhya Pradesh. The tin slags here contain significant amounts of Ta (upto 20% Ta_2O_5).

Fluorite-bearing peraluminous granites at Kanigiri, Prakasam district, Andhra Pradesh along the eastern margin of Cuddapah basin have been found to be enriched in Nb, Ta, Sn, U, Th and Rb. Discrete phases of columbite and samarskite occur in them along with ilmenite, magnetite, zircon, monazite, garnet and rutile (Banerjee, 1999).

Abundant deposits of RM and REE minerals of columbite-tentalite, beryl, lepid-olite, samarskite and fergusonite have been reported from the pegmatites occurring along Bihar Mica Belt. Apart from mica, this belt has a commercial potential for columbite-tentalite and beryl. Several tons of beryl were recovered by private mine owners, while mining for mica, and the same was supplied to AMD. Similarly, REE and Rare Metal (Nb, Li, Cs) are significantly enriched in rocks of alkaline-carbonatite complex of South and North Purulia Shear Zone as detailed in Chapter 6. Genetically the mineralization is connected with subcrustal alkaline magmatism. The post magmatitic mineralization of Rare Earths is found within a wide range of deposits. In addition to REE, niobium mineralization with no significant tantalum occurs in alkaline rocks of Purulia district.

Exploration and exploitation of Rare Metal pegmatities during the past fifty years have made India self sufficient in terms of Nb, Ta, Be and Li resources that are required for the nuclear power programmes and to a limited extent for other high technology applications. Several rare minerals such as beryl, columbite-tentalite, lepidolite and spodumene have been produced either alone or as co-product from a number of pegmatite belts notably from Bihar, Rajasthan, Madhya Pradesh and Karnataka. However, considering the likely increase in demands in future for a vari-ety of rare metals and rare earths, it is imperative that a vigourous integrated and holistic approach by using modelling and geochemical exploration techniques should be taken up in continental areas. Similar approaches on stream sediments and coastal areas may be followed up along the similar terrains of any country with conceptual approaches, which would be necessary to discover new resources of any country of the world.

In this connection mention may be made about the application of Zipf's Law (1949) for resource prediction. In this law a mathematical relationship between size and rank of discrete phenomenon are used for resource prediction as already detailed earlier. Paliwal et al. (1986) and Halder (2004) applied this model to Pb-Zn metal accumulation of 24 known Precambrian deposits in India, and according to them about 75% Pb-Zn metal is yet to be discovered in India and more effort should be given in exploring the Banded Gneissic Complex of Rajasthan and other similar terrains. Application of Zipf's law may be applied for the other commodities also in different parts of the terrain, to predict the chance of locating new mineral deposits of India, as well as different concealed mineral deposits of Asian countries and other countries of the world. In this connection whole hearted effort should be made for discovering the concealed mineral deposits, and one should be hopeful while trying to locate such deposits.

Appendix

Table A1. Chemical Symbols and Elements

Ac	Actinium	Ge	Germanium	Pr	Praseodymium
Ag	Silver	H	Hydrogen	Pt	Platinum
Al	Aluminium	He	Helium	Pu	Plutonium
Am	Americium	Hf	Hafnium	Ra	Radium
Ar	Argon	Hg	Mercury	Rb	Rubidium
As	Arsenic	Ho	Holonium	Re	Rhenium
At	Astatine	I	Iodine	Rh	Rhodium
Au	Gold	In	Indium	Rn	Radon
B	Boron	Ir	Iridium	Ru	Ruthenium
Ba	Barium	K	Potassium	S	Sulphur
Be	Beryllium	Kr	Krypton	Sb	Antimony
Bi	Bismuth	La	Lanthanium	Sc	Scandium
Bk	Berkelium	Li	Lithium	Se	Selenium
Br	Bromine	Lr	Lawrencium	Si	Silicon
C	Carbon	Lu	Lutetium	Sm	Samarium
Ca	Calcium	Md	Mendelevium	Sn	Tin
Cd	Cadmium	Mg	Magnesium	Sr	Strontium
Ce	Cerium	Mn	Manganese	Ta	Tantalum
Cf	Californium	Mo	Molybdenum	Tb	Terbium
Cl	Chlorine	N	Nitrogen	Tc	Technetium
Cm	Curium	Na	Sodium	Te	Tellurium
Co	Cobalt	Nb	Niobium	Th	Thorium
Cr	Chromium	Nd	Neodymium	Ti	Titanium
Cs	Caesium	Ne	Neon	Tl	Thallium
Cu	Copper	Ni	Nickel	Tm	Thulium
Dy	Dysprosium	No	Nobelium	U	Uranium

(continued)

© Capital Publishing Company, New Delhi, India 2020
A. K. Talapatra, *Geochemical Exploration and Modelling of Concealed Mineral Deposits*, https://doi.org/10.1007/978-3-030-48756-0

Table A1. (continued)

Es	Einsteinium	Np	Neptunium	V	Vanadium
Er	Erbium	O	Oxygen	W	Tungsten
Eu	Europium	Os	Osmium	Xe	Xenon
F	Fluorine	P	Phosphorus	Y	Yttrium
Fe	Iron	Pa	Protactinium	Yb	Ytterbium
Fm	Fermium	Pb	Lead	Zn	Zinc
Fr	Francium	Pd	Palladium	Zr	Zirconium
Ga	Gallium	Pm	Promethium		
Gd	Gadolinium	Po	Polonium		

Table A2. Abundance of Elements in Average Crustal Rocks (ppm)

Element	Abundance	Element	Abundance
Aluminium	81,000	Mercury	0.02
Antimony	0.1	Molybdenum	1.5
Arsenic	2	Nickel	75
Barium	580	Niobium	20
Beryllium	2	Oxygen	473,000
Bismuth	0.1	Palladium	0.01
Boron	8	Phosphorus	900
Bromine	1.8	Platinum	0.005
Cadmium	0.1	Potassium	25,000
Calcium	33,000	Rhenium	0.0006
Carbon	230	Rubidium	150
Cerium	81	Scandium	13
Cesium	3	Selenium	0.1
Chlorine	130	Silicon	291,000
Chromium	100	Silver	0.05
Cobalt	25	Sodium	25,000
Copper	50	Strontium	300
Fluorine	600	Sulphur	300
Gallium	26	Tantalum	2
Germanium	2	Tellurium	0.002
Gold	0.003	Thallium	0.45
Hafnium	3	Thorium	10
Indium	0.1	Tin	2
Iodine	0.15	Titanium	4400
Iron	46,500	Tungsten	1
Lanthanum	25	Uranium	2.5
Lead	10	Vanadium	150
Lithium	30	Zinc	80
Magnesium	17,300	Zirconium	150
Manganese	1000		

Source: Levinson (1974). Appendix data (average of contents in granite and mafic rocks)

Table A3. Conversion Table

Length	**Currency**
1 nautical mile = 1.85 km	1 lakh = 0.1 million
1 mile = 1.609 km	1 crore = 100 lakhs = 10 million
1 km = 0.621 mile	1 billion = 1000 million
1 km = 1000 m = 1,00,000 cm = 1,000,000 mm	
1 mm = 1000 μm (micrometre or microns)	
Volume	**Mass**
1 litre = 1000 millilitres = 0.264 gallons	1 kilogram = 1000 grams = 2.205
1 gallon = 3.785 litres	pounds
1 barrel = 42 gallons	1 pound = 0.454 kilogram
Size and Concentration	
Million = 1,000,000 = 10^6	1 ppm = micro G/G = 10^{-6} g/g
Trillion = 1,000,000,000,000 = 10^{12}	1 ppb = nano G/G = 10^{-9} g/g
1 Micro = 1/million = $1/10^6$ = 1×10^{-6}	1 ppt = pico G/G = 10^{-12} g/g
1 Nano = 1/billion = $1/10^9$ = 1×10^{-9}	1 ppm $\times 10^3$ = 1 ppb
1 Pico = 1/trillion = $1/10^{12}$ = 1×10^{-12}	1 ppm $\times 10^6$ = 1 ppt
1 Fico = $1/10^{15}$ = 1×10^{-15}	1 ppb $\times 10^{-3}$ = 1 ppm
For waters (dilute solutions)	**For gases (atmosphere, soils, exhalation)**
1 mg/l = 10^{-3} g/l = 1000 ppb or 1 ppm	1 ng/m³ = 10^{-3} ng/l
$\quad\quad 10^{-4}$ g/l = 100 ppb	1 ng/m³ = 10^{-6} μg/l
$\quad\quad 10^{-5}$ g /l = 10 ppb	1 ng/m³ = 10^{-9} mg/l
1 μg/l = 10^{-6} g /l = 1 ppb	1 ng/m³ = 10^{-12} g/l
	100 ng/m³ = 10^{-7} mg/l

Table A4. The Geological Time Scale

Era			Period	Epoch	Ended- began (Ma)
Cenozoic		Quaternary		Holocene	0–0.018
				Pleistocene	0.018–1.6
	Tertiary		Neogene	Pliocene	1.6–5.1
				Miocene	5.1–24
			Paleogene	Oligocene	24–38
				Eocene	38–55
				Paleocene	55–65
Mesozoic	Cretaceous				65–144
	Jurassic				144–213
	Triassic				213–248
Paleozoic	Permian				248–286
	Carboniferous				286–360
	Devonian				360–408
	Silurian				408–438
	Ordovician				438–505
	Cambrian				505–570
Proterozoic					570–2700
Archaean					2700–4000
Hadean					4000–4560

Table A5. Magma Composition in Relation to Various Plate Tectonic Settings

	Plate margin		Intraplate area	
Plate setting	Converging (Subduction zone)	Diverging (Oceanic Ridge)	Marginal sea basin	Ocean Island and Seamounts
Magma type	Calc-alkaline (tholeiite)	Tholeiite (low K)	Low K Tholeiite (Calc-alkaline/	Tholeiite (Alkali)
Stress regime	Compressive	Extension	Extension	Minor compressive

Source: Condie, 1982; Under magma type, the variety in parenthesis is secondary in abundance.

Table A6. Geochemical Classification of Elements

Type	Element
Siderophile	Fe, Co, Ni, Pd, Au, Ge, Ga, Cu, As, Sb, Mo
Chalcophile	Ag, Zn, Cd, Hg, Sn, Pb, Bi
Lithophile	Li, Na, K, Rb, Cs, Be, Mg, Ca, Sn, Ba, Al, La, Lu, Si, Ti, Zr, P, V, Nb, H, F, Cr, Br, I
Atmophile	(H), N, (O), He, Ne, Ar, Kr, Xe

Source: Mason, 1966

Table A7. Paleo-Sedimentation Rate and Sediment Thickness in the World Oceans

Geological age	Length of period (Ma)	Sedimentation rate (km/Ma)	Sediment thickness (km)
Neogene and Quarternary	26	0.50	13
Paleogene	39	0.54	20.9
Cretaceous	71	0.22	15.8
Jurassic	54	0.24	13.1
Triassic	35	0.25	8.8
Permian	55	0.11	6.2
Carboniferous	65	0.21	13.8
Devonian	50	0.23	11.7
Silurian	35	0.25	8.9
Ordovician	70	0.20	13.8
Cambrian	70	0.17	11.8
Total	570	0.26 (av.)	137.8

Note: Present day sedimentation rate (mm/10^3 yr) is about 5.8 in the Atlantic Ocean, 4.4 in the Indian Ocean and 3.4 in the Pacific Ocean (Ghosh and Mukhopadhyay, 1999)

Table A8. Size Classification of Sediments

Classification	Size (mm)	Sinking rate (cm/s)
Bolder	> 256	$>4.29 \times 10^6$
Cobble	64-256	$2.68 \times 10^5 - 4.29 \times 10^6$
Pebble	4-64	$1.05 \times 10^3 - 2.68 \times 10^5$
Granule	2-4	$2.62 \times 10^2 - 1.05 \times 10^3$
Coarse – Medium sand	0.25-2	4.09-262
Fine sand	0.0625-0.25	0.256-4.09
Silt	0.0039-0.0625	$9.96 \times 10^{-4} - 2.56 \times 10^{-1}$
Clay	<0.0039	$<9.96 \times 10^{-4}$

References

Agterberg, F.P. (1974). Geomathematics – Mathematical background and geoscience application. Elsevier, Amsterdam, p. 594.

Agterberg, F.P., Chung, C.F., Fabbri, A.G., Kelly, A.M. and Spranger, J.S. (1972). Geomathematical evaluation of copper and zinc potential of the Abitibi area Ontario and Quebec. Geol. Survey of Canada Paper **71–41:** 1–55.

Anon (1976). Mercury vapour trap for exploration of concealed base metal mineralization. *Geol Survey of India News*, **7(4):** 9.

Armour-Brown, A. and Nichol, I. (1970). Regional geochemical reconnaissance and the location of metallic provinces. *Econ. Geol.*, **65:** 312–330.

Bandopadhaya, S., Sengupta, D. and Sengupta, R. (2002). Computer aided graphic representation of heavy mineral placer data and generation of thematic maps along the coastal areas. GSI unpublished Report.

Banerji, P.K. (1983). A morphotectonic classification of laterite and ironstone covered basements in Gondwana land – A case study from Orissa, India. Proc. 2nd Int. Sem. Late. Process, Sao Paolo (Eds. A.J. Melfi and A. Carvalho). 371–400.

Banerji, P.K. (1989). Understanding the regolithic landform for identifying exploration targets for strategic, rare and precious metal ore deposits in East and South Indian Shield. Seminar on Planning for mineral exploration by 2000 A.D. Mining, Geological and Metallurgical Institute of India, Calcutta.

Banerji, P.K., Ghosh, S., Bose, P., Sengupta, P.K., Kar, S.K., Kar Roy, M.K., Mitra, F.K. and Ghose, B.K. (1982). Programs and methodology of integrated survey. GSI Misc. Publ. No. 8.

Banerjee, D.C. (1999). Rare Metal and Rare Earth Pegmatites of India: An overview and some perspective. Special Issue on 'Rare Metal and Rare Earth Pegmatites of India'. **12:** 1–6.

Banerjee, A.K. and Talapatra, A.K. (1966). Soda-granite from South of Tatanagar, Bihar, India. Geol Mag. (U.K.), **103:** 342–351.

Banerjee, D.C., Ranganath, N., Maithani, P.B. and Jayaram, K.M.V. (1987). Rare metal bearing pegmatites in parts of southern Karnataka, India. *Journal of Geological Society of India*, **30:** 507–513.

Banerjee, D.C., Krishna, K.V.G., Murthy, G.V.G.K., Srivastava, S.K. and Sinha, R.P. (1994). Occurrence of Spodumene in the rare metal bearing pegmatites of Marlagalla-Allapatna area, Mandya district, Karnataka. *Journal of the Geological Society of India*, **44:** 127–139.

Banerjee, Buddhadeb, De, S.S. and Das, L.K. (1992). Exploration of Metallic Minerals and their discrimination using partial extraction of metals by Electrolysis (PEXMEL) Method. *Jour. Geol. Soc. of India*, **40:** 151–161.

© Capital Publishing Company, New Delhi, India 2021
A. K. Talapatra, *Geochemical Exploration and Modelling of Concealed Mineral Deposits*, https://doi.org/10.1007/978-3-030-48756-0

Benest, J.M. and Winter, P.E. (1984). Ore reserve estimation by use of geologically controlled geostatistics. 18th Int. Sym., APCOM' **84:** 367–388.

Bhadra Chaudhury, J.N., Helfmeier, H., Chakravorty, P.S., Newsely, H. and Mucke, A. (1989). Notes on occurrence of Economically Important REE-bearing Minerals in Pegmatites of Northern shear of Purulia District, West Bengal, India. Abstr. 28th IGC, **1:** 146

Bhattacharya, C., Talapatra, A.K. and Bose, S.S. (1984). Integrated geological approach for tracing gold mineralisation in parts of Singbhum and Ranchi districts, Bihar. GSI Record No. 114, Part II: 1–14.

Biswas, P.K. (1997). Statistical and Geostatistical modelling of the Mine Block, Rakha Copper Deposits. Ph.D. Thesis, ISM Dhanbad.

Bonham-Carter, G.F., Agterberg, F.P. and Wright, D.F. (1989). Weights of evidence modelling: A new approach to mapping mineral potential. *Geological Survey of Canada Paper*, **89(9):** 171–183.

Botbol, J.M. (1971). An application of characteristics analysis to mineral exploration. *Canadian Inst. Min. and Metall.*, Spec. v. **12:** 92–99.

Brooks, R.R. (1972). Geobotany and Biochemistry in Mineral Exploration. Harper and Row. New York, p. 290

Brown, J.C. and Dey, A.K. (1955). India's mineral wealth. Oxford University Press, p. 761.

Burke, K., Dewey, J.F. and Kidd, W.S.F. (1976). Dominance of horizontal movements, arc and microcontinental collision during the later permobile regime. *In:* The Early History of the Earth. Windley, B.F. (Ed.), pp. 113–129.

Butt, C.R.M. (1981). Some aspects of geochemical exploration in lateritic terrains in Australia. *In:* Lateritisation Processes, Proceedings of the Inter. Seminar on Lateritisation Processes, pp. 369–380.

Butt, C.R.M. and Smith, R.E. (1980). Conceptual Models in Exploration Geochemistry: *Australia. Jour. Geochem. Expl.*, **12:** 89–365.

Cameron, E.M. (1977). Geochemical dispersion in mineralized soils of a permafrost environment. *Jour. Geochem. Explor.*, **7:** 301–326.

Cameron, E.M. and Jonasson, I.R. (1972). Mercury in Precambrian shales of the Canadian shield. *Geochim et Cosmochim Acta.*, **36(9):** 985–1006.

Cerny, P. (1991). Fertile granites of Precambrian Pegmatite fields; its Geochemistry controlled by Techtonic setting or Source Lithologies. *Precambrian Research*, **51:** 429–468.

Chaffee, M.A. (1976). Geochemical exploration techniques based on distribution of selected elements in rocks, soils, and plants, Mineral Butte Copper deposit, Pinal county, Arizona. *U.S. Geol. Surv. Bull.*, **1278-D:** 55.

Chaffee, M.A. and Gale, C.W. (1976). The California poppy (*Eschsholtzia mexicana*) as a copper indicator plant – A new example. *Jour. Geochem. Explor.*, **5:** 59–63.

Chilov, S. (1975). Determination of small amount of mercury – A review. *Talanta*, **22:** 205–232.

Clark, I. (1979). Practical Geostatistics. Academic Press, London, p. 129.

Colliyer, P.L. and Merrium D.F. (1973). An application of cluster analysis in mineral exploration. *Math. Geol.*, **5(3):** 213–223.

Condie, K.C. (1982). Plate Tectonics and crustal evolution. Pergamon Press Inc. New York, p. 310.

Das Gupta, S.P. (1964). Genesis of sulphide mineralisation in Khetri copper belt, Rajasthan, India. Report 22nd Int. Geol. Congr. Pt., **5:** 239–256.

Das Gupta, S.P. (1968). Structural history of the Khetri copper belt, Jhunjhunu and Sikar Dist., Rajasthan. Mem. Geol. Surv. Ind., **98:** 170.

Das Gupta, S.P. (1974). Geological setting and origin of sulphide deposits in the Khetri Copper Belt, Rajasthan. Geol. Min. Met. Soc. Ind. **Golden Jubilee Volume:** 223–238.

David, M. (1980). Grade tonnage problems. *In:* Computer Methods for the 80's, ed. A. Weiss, APCOM-80, **Section 2:** 171–189.

David, M., Dimitrikopoulas, R. and Marcottes, D. (1987). GEOSTAT 1 – A prototype expert system for explicit knowledge approach to geostatistics. Proc. 20th APCOM, Johannesburg, **3:** 121–126.

Davidson, C.F. (1956). The Economic Geology of Thorium. *Mining Mag.*, **94(4):** 197–208.

Davis, John C. (1973). Statistics and Data Analysis in Geology. John Wiley & Sons. New York, p. 550.

Davis, B.M. and Borgman, L.E. (1979). A test of hypothesis concerning a proposed model for the underlying variogram. Proc. 16th APCOM, pp. 163–181.

Davy, R. and Stokes, M. (1976). A preliminary evaluation of the use of sulphur dioxide as a prospecting tool in W. Australia. West. Aust. Geol. Surv. Geochem. Report, **8:** 72–76.

Deb, M. and Pal, T. (2004). Geology and genesis of base metal sulphide deposits in the Dariba-Rajpura-Bethumni Belt, Rajasthan India in the light of basin evolution. *In:* Deb, M. and Goodfellow, W.D. (Eds.), Sediment hosted lead-zinc sulphide deposits: Attributes and models of some major deposits in India, Australia and Canada. Narosa Publishing House. New Delhi, pp. 304–327.

Dimitri-kopoulas, R. (1993). Artificially intelligent Geostatistics – A framework accommodating qualitative knowledge information. *Mathematical Geology*, **25(3):** 261–279.

Doe, B.R. and Stacey, J.S. (1974). The application of the lead isotopes in the problem of ore genesis and ore prospect evaluation: A review. *Econ. Geol.*, **67:** 757–776.

Donn, W.L., Farrand, N.L. and Ewing M. (1962). Pleistocene ice volume and sea level lowering. *Jour. Geol.*, 206–214.

Elliot, I.L. and Fletcher, W.K. (Eds.) (1974). Geochemical Exploration: Developments in Economic Geology. Elsevier Sci. Pub. Co.

Fletcher, W.K. (1981). Analytical methods in Geochemical Prospecting. Elsevier, Amsterdam, p. 255.

Garrels, R.M. and Christ, C.L. (1965). Solutions, minerals and equilibrium. Harper and Row, New York, p. 450.

Garret, R.G. and Nichol, I. (1967). Regional Geochemical Reconnaissance in Eastern Sierra Leone. *Trans. Inst. Min. Metall., Sect. B*, **76:** 97–112.

Ghose, N.C. (1992). Chotonagpur gneiss-granite complex, Eastern India—Present status and future prospect. *Ind. Jour. of Geology*, **64:** 100–121.

Ghose, A.K. and Mukhopadhyay, R. (1999). Mineral wealth of the ocean. Oxford & IBH Publishing Co., p. 225.

Ginsberg, I.T. (1960). Principles of Geochemical Prospecting. Pergamon Press.

Goodenough, K. M., Wall, Frances and Merriman, David (2017). The Rare Earth Elements: demands, Global resourcing and challenges for resourcing future generations. *Natural Resources Research*, **27(2):** 201–216.

Goodwin, A.M. (1977). Archaean basin – Craton complexes and the Precambrian shields. *Canad. Jour. Earth Sci.*, **14.**

Goodwin, A.M. (1981). Archaean plates and greenstone belts. *In:* Precambrian Plate Tectonics. Elsevier, Amsterdam.

Gordon, G.E., Romdle, K. and Goles, G.G. (1968). Instrumental neutron activation analysis of standard rocks with high resolution gamma ray detectors. *Geochim. Cosmochim.Acta.*, **32:** 369–396.

Govett, G.J.S. (Ed.) (1983). Handbook of Exploration Geochemistry. Vol. 2. Elsevier Sci. Publ. Co., p. 437.

Green, G.R., Ohmoto, H., Date, J. and Takahashi, T. (1983). Whole rock oxygen isotope distribution in the Fukazawa-Kosaka area, Hokuroku district, Japan, and its potential application to mineral exploration. *Econ. Geol.* (Monograph), **5:** 395–411.

Gulson, B.L. (1977). Application of lead isotopes and trace elements to mapping black shales around a base metal sulphide deposit. *Jour. Geochem. Explo*, **8:** 85–104.

Halder, S.K. (2004). Grade and tonnage relationships in sediment hosted lead zinc sulphide deposits of Rajasthan, India. *In:* Deb, M. and Goodfellow, W.D. (Eds.), Sediment hosted lead-zinc sulphide deposits: Attributes and models of some major deposits in India, Australia and Canada. Narosa Publishing House. New Delhi, pp. 264–272.

Hawkes, H.E. and Webb, J.S. (1962). Geochemistry in Mineral Exploration. Harper and Row Publishers. New York.

Hawkes, H.E. (1972). Exploration Geochemistry Bibliography, 1965-1971. Spec. Vol. No. 1, Assoc. Expl. Geochem., Toronto, p. 118.

Hawkes, H.E. (1976). Exploration Geochemistry Bibliography, 1972-1975. Spec. Vol. No. 5, Assoc. Expl. Geochem., Toronto, p. 195.

Hood, P.J. (Ed.) (1979). Geophysics and Geochemistry in the search for metallic ores. Geol. Surv. Can., Econ. Geol. Rep., **31:** 811.

Henderson, P. (2013). Rare Earth Element Geochemistry. Elsevier.

Jayaram, B.M. and Ravindran, K.V. (1979). Copper deposits in Aladahalli area, Hassan District, Karnataka. *Indian Miner.*, **33(3):** 33–44.

Jhonson, K.R. (1995). Geological modelling – The way ahead: Concept of modelling for exploration and mine planning. Proc. 25th APCOM, Brisbane, pp. 1–4.

Jonasson, I.R. and Boyle, R.W. (1972). Geochemistry of mercury and origins of natural contamination of the environment. *CIM Transactions*, **75:** 8–15.

Jordens, A., Cheng, Y.P. and Waters, K.E. (2013). A review of the beneficiation of rare earth element bearing minerals. *Minerals Engineering*, **41:** 97–114.

Kesler, S.E. (1973). Copper, molybdenum and gold abundances in porphyry copper deposits. *Econ. Geol.*, **68:** 106–113.

Koch Jr., G.S. and Link, R.F. (1970). Statistical analysis of Geological data. Vol. I & II, John Wiley & Sons. Inc., New York, p. 375 and p. 438.

Kroner, A. (1981). Precambrian plate tectonics. *In:* Kroner (Ed.), Precambrian Plate Tectonics. Elsevier, Amsterdam, pp. 350–447.

Krishnamurthy, P (1988). Carbonatites of India—Exploration and Research for Atomic Minerals, Vol. 81. AMD, DAE, Govt. of India.

Krishnamurthy, N. and Gupta, C.K. (2015). Extractive Metallurgy of Rare Earths (2nd ed.). Boca Raton: CRC Press.

Krumbein, W.C. and Graybill, F.A. (1965). An Introduction to Statistical Models in Geology. McGraw Hill, p. 475.

Kumari Sandhya (1996). Geostatistical Modelling and Resource Evaluation of Bauxite Deposit, Durgamanwadi Mines, Kolhapur District, Maharashtra. M. Tech. Thesis. ISM, Dhanbad.

Kumari, A., Panda, R., Jha, M.K., Kumar, J.R. and Lee, J.Y. (2015). Process development to recover rare earth metals from monazite mineral: A review. *Minerals Engineering*, **79:** 102–115.

Kunzendorf, H. (1986). Marine mineral explorations. Elsevier, Amsterdam, p. 300.

Lambert, I.B. and Groves, D.I. (1981). Early earth evolution and metallogeny. *In:* K.H. Wolf (Ed.), Handbook of Stratabound and Stratiform Ore Deposits. Elsevier, Amsterdam, 350–447.

Lepeltier, C. (1969). A simplified statistical treatment of geochemical data by graphical representation. *Econ. Geol.*, **64:** 538–550.

Levinson, A.A. (1974). Introduction to Exploration Geochemistry. Applied Publishing Ltd.

Levinson, A.A. and de Pablo, L. (1975). A rapid X-ray fluorescence procedure applicable in exploration geochemistry. *Jour. Geochem. Expl.*, **4:** 399–408.

Lovering, T.G. and McCarthy, J.H. Jr. (1978). The basin and Range Province of the Western United States and Northern Mexico. *Jour. Geochem. Expl.*, **9:** 143–162.

Mahadevan, C. and Sriram Das, A. (1948). Monazite in the sands of Vishakhapatnam District. *Proc. Ind. Acad. Sci.*, **27A:** 273–278.

Mallick, T.K. (1968). Heavy minerals of shelf sediments between Vishakhapatnam and the Penner Delta, eastern coast of India. *Bull. Nat. Ins. of Sci. Ind.*, **38:** 502–512.

Mallick, T.K. (1972). Opaque minerals from the shelf sediment of Mangalore, western coast, India. *Mar.Geol.*, **12:** 207–222.

Mallick, T.K. (1974). Heavy mineral placers in the beaches and offshore areas—Their nature, origin, economic potential and exploration. *Indian Minerals*, **28(3):** 39–46.

Mallick, T.K. (1981). Distribution patterns of heavy minerals from the northern part of the Godavari Delta of Kakinada, A.P. *Ind. Jour. Mar. Sci.*, **10(1):** 51–56.

Mallick, T.K. (1983). Shelf sediments and mineral distribution patterns off Mandapam, Palk Bay. *Ind. Jour. Mar. Sci.*, **12**: 203–208.

Mallick, T.K. (1986). Micro morphology of some placer minerals from Kerala Beach, India. *Marine Geology*, **71**: 371–381.

Mallick, T.K. (2002). Exploration of Mineral Resources around Indian Coast. Proceedings of the National Conference on Exploration, Mining and Processing of beach Placers in India. N. Chandrasekhar and P. Sivasubramaniam (Editors), pp. 56–78.

Mallick, T.K. (2007). Mineral resources of the beaches and shallow seas around Indian Coast. Proc. Workshop on Exploration and Evaluation of Coastal and Offshore Resources. *Bulletin I.G.C.*, **1(2)**: 80–87 (Edited by Prof. O.P. Verma).

Mallick, T.K. (2015). Resources of the near-coastal and offshore sediments. *Transactions Mining Geological and Metallurgical Institute of India*, **3**: 9–25.

Mallick, T.K. and Sen Sharma, Kakoli (2009). Highlights of heavy mineral distribution patterns along the coastline of West Bengal. *Ind. Jour. of Geosciences*, **63(4)**: 429–442.

Mallick, T.K., Venkatesh, K.V., Sengupta, R., Rao, B.R. and Rama Murthy, M. (1976). Heavy mineral distribution patterns in the northern part of the Arabian Sea, Western Continental Shelf of India. *Indian Journal of Earth Science*, **3(2)**: 178–187.

Mallick, T.K., Vasudevan, V., Aby Verghese, P. and Machado, T. (1987). The black sand placer deposits of Kerala. *Marine Geology*, **77**: 129–150.

Massari, S. and Ruberti, M. (2013). Rare earth elements as critical raw materials: Focus on international markets and future strategies. *Resources Policy*, **38(1)**: 36–43.

Mason, B. (1966). Principles of Geochemistry. John Wiley & Sons. New York, p. 329.

Matheron, G. (1971). The theory of regionalized variables and its application. Booklet No. 5, Les Cahiers du Centre de Morphologie Mathematique, Fontainebleau, p. 211.

Mazumdar, S.K. (1988). Crustal evolution of Chotanagpur gneissic complex and the Mica Belt of Bihar. *In:* Precambrian of the Eastern Indian Shield. D. Mukhopadhyay, (Ed.). *Geol. Soc. Ind. Min.*, **8**: 49–83.

McCammon, R.B. (1986). The PROSPECTOR Mineral consultant system. USGS Bull., 1697, p. 35.

McCarthy, J., Vaughn, W.W., Learned, R.E. and Meuschke, J.L. (1969). Mercury in soil-gas and air—A potential tool in mineral exploration. U.S. Geol. Surv. Cir. 609, p. 16.

McCarthy, J.H. Jr. (1972). Mercury vapour and other volatile components in the air as guide to ore deposits. *Jour. Geochem. Expl.*, **1**: 143–162.

McNerney, J.J. and Buseek, P.R. (1973). Geochemical exploration using mercury vapour. *Econ. Geol.*, **68**: 1313–1320.

Mookherjee, A. (1999). Ore genesis – A holistic approach. Allied Publishers, Ltd., p. 658.

Mukherji, B. (1996). Deposit Modelling: A tool for predicting unknown mineral deposits. Workshop on Mineral Deposit Modelling. ISM, Dhanbad, **Abstact v:** 57.

Mukherji, B. (2003). Geospatial modelling of copper mineralization in Singhbhum belt, Jharkhand. Ph.D. Thesis. ISM Dhanbad.

Mukhopadhyay, D. (1989). Precambrian plate tectonics in the Eastern Indian Shield. *In:* Crustal Evolution and Orogeny. S.P.H. Sychanthrong (Ed.). Oxford & I.B.H. Publishing, New Delhi, pp. 75–100.

Nassar, N.T., Du, X. and Graedel, T.E. (2015). Criticality of the rare earth elements. *Journal of Industrial Ecology*, **19(6)**: 1044–1054.

Nichol, I., James, L.D. and Viewing, K.K. (1966). Regional geochemical reconnaissance in Sierra Leone. *Trans. Inst. Mining and Metall.*, London (Sect. B), **75**: 146–161.

Numbiar, A.R. and Unnikrishnan, E. (1989). Distribution and origin of heavy minerals in the recent shelf sediments off Trivandrum-Muttam coast and adjacent beaches, south west India. *Geol. Surv. Ind.*, Spl. Pub., **24**: 247–255.

Ohmoto, H. (1986). Stable isotope geochemistry of ore deposits. *In:* Stable Isotopes in high temperature geologic processes. John W. Valley, Hugh P. Taylor, Jr and James R. O'Neil (Eds). Review in Mineralogy, Vol. 16, Min. Soc. Amer. Washington, D.C., pp. 491–559.

Ohmoto, H. and Rai, R.O. (1979). Isotopes of sulphur and carbon. *In:* Geochemistry of Hydrothermal Ore Deposits, 2nd edition. H.L. Barnes (Ed.). John Wiley & Sons. New York, pp. 509–567.

Overstreet, W.C. and Olson, J.C. (1964). Geologic distribution and resources of thorium. U.S. Geological Surv. Bulletin. **1264:** 61.

Ozerova, N.A. (1962). Primary dispersion haloes of mercury. Proc. Inst. Geol. Ore Deposits; Petrogr. Mineralog. and Geochem. No. 72, Questions of geochemistry, Part 4. Nauka Press, Moscow. (Transl. Internat. Geol. Rev. 1970).

Paliwal, H.V., Bhatnagar, S.N. and Haldar, S.K. (1986). Lead-zinc resource prediction in India: An application of Zipf's Law. *Mathematical Geology*, **18(6):** 539–549.

Pirajno, F. (2000). Ore Deposits and Mantle Plume. Kluwer Academic Publishers, p. 56.

Plant, J.A., Hale, M. and Ridgway, J. (1988). Developments in regional geochemistry for mineral exploration. *Trans. Inst. Min. Metall., Sec. B – Applied Earth Science*, **97:** B116–B140.

Podder, B.C. (1972). Base metal mineralisation in the Rajpura belt, Udaipur district, Rajasthan. Record Geol. Surv. India, **104 (Pt. 1):** 16–19.

Poddar, B.C. (1974). Evaluation of sedimentary sulphide rhythmites into metamorphic tectonites in the base-metal deposit of Rajpura, Rajasthan. Geological Mining and Metallurgical Society of India, **Golden Jubilee Volume:** 207–222.

Poddar, B.C. and Chatterjee, A.K. (1966). An interim report on the Dariba-Rajpura belt, Udaipur Dist. Rajasthan. Unpublished Report, Geol. Surv. Ind.

Poddar, B.C. and Mathur, R.K. (1963). A preliminary note on the Dariba-Rajpura belt, Udaipur dist., Rajasthan. Unpublished Report. G.S.I.

Polynov, B.B. (1937). The Cycle of Weathering. (A. Muir, Translator) Murby, London, p. 220.

Prasad, E.A.V. (1987). Geobotany and biogeochemistry in mineral exploration in the tropics. *Jour. Geochem. Expl.*, **29:** 427–428.

Rai, S.D., Shivananda, S.R., Tiwari, K.N., Banerjee, D.C. and Kaul, R. (1991). Xenotime-bearing inland placers in India and their benification, exploration and research for atomic minerals. AMD, DAE, Govt. India, **4:** 47–92.

Raja Rao, C.S. and Chatterjee, A.K. (1972). Dariba-Rajpura-Bethumi Belt of zinc-lead mineralisation, Udaipur District, Rajasthan. Geological Survey of India, Miscellaneous Publication **16:** 617–626.

Raja Rao, C.S., Mathur, R.K., Dhara, M.K. and Poddar, B.C. (1970). Report on the exploration for zinc-lead at Rajpura-Dariba Prospect, Udaipur District, Rajasthan. Unpublished Report, G.S.I.

Ramesh Babu, P.V., Pandey, B.K. and Dhana Raju, R. (1993). Rb-Sr ages on the granite and pegmatite minerals from Bastar-Koraput belt, Madhya Pradesh and Orissa, India. *Journal of Geological Society of India*, **42:** 33–38.

Rao, B.R.J. and Shrivastava, P.C. (1996). Non-living resources of the exclusive economic zone of India. *Ind. Minerals*, **50(1&2):** 53–62.

Rao, B.R.J,, Mahapatra, G.P., Vaz, G.G., Reddy, D.R.S, Hariprasad, M., Misra, U.S., Raju, D.C.L. and Shankar, J. (1992). Inner shelf placer sands off north Andhra coast. *Marine Wing News Letter*, **8(1):** 11–22.

Ravi Shankar (2001). Geodynamic processes and metallogeny. GSI Sp. Publ. No. **72:** 1–10.

Ravindran, K.V. (1982). The final report on exploration for base metal in the Aladahalli Schist Belt, Hassan district, Karnataka. Unpublished G.S.I. Report.

Ray, D.K., Banerjee, P.K., Pant, A., Sinha Roy, S., Ramkrishna, M., Saha, A.B., Raghunandan, K.R. and Choudhury, N.P. (1987). Report on 'Strategy for concept oriented mineral exploration (GOS-COMINEX). Geological Survey of India.

Reedman, J.H. (1979). Techniques in Mineral Exploration. Applied Science Publication.

Rendu, J.M. (1981). An Introduction to Geostatistical Methods of Mineral Evaluation. Monograph SAIMM, Johannesburg, p. 84.

Roonwall, G.S. (1986). The Indian Ocean: Exploitable mineral and petroleum resources. Narosa Publishing House, p. 198.

Roedder, E. (1971). Fluid inclusion studies on the porphyry type ore deposits at Bingham, Utah, Butte, Montana and Climax Colorado. *Econ. Geol.*, **66**: 98–120.

Rose, A.W., Dahlberg, E.C. and Keith, M.L. (1970). A multiple regression technique for adjusting background values in stream sediment geochemistry. *Econ. Geol.*, **65**: 156–165.

Rose, A.W., Hawkes, H.E. and Webb, J.S. (1979). Geochemistry in Mineral Exploration. Academic Press, p. 657.

Roskill (2016). Rare earths: Global industry, markets and outlook (16th ed.). London, UK: Roskill.

Rouse, C.E. and Stevens, D.E. (1971). The use of the sulphur dioxide geochemistry in the detection of sulphide deposits. Assoc. Met. Inst. Min. Eng.

Roy Chowdhury, M.K., Das Gupta, S.P., Prasad Rao, G.H.S.V., Venkatesh, V., Ramaiengar, A.S. and Natarajan, W.K. (1968). Geologic potential of the Khetri Copper Belt, Rajasthan. Misc. Publ. G.S.I. No. **13**: 165–182.

Saha, A.K. and Talapatra, A.K. (1962). An occurrence of Gabbro-granophyre association of rocks near Jorapokhar in Singhbhum District, *Bihar. Science & Culture*, **28**: 582–585.

Saha, A.K., Sarkar, S.N., Basu, Swapna and Ganguly, D. (1986). A multivariate statistical study of copper mineralisation in the central section of Mosaboni mine, Eastern Singhbhum, India. *Math. Geol.*, **18(2)**: 215–235.

Sarkar, B.C. (2005). Developments in geomathetical modelling and computer application in mineral resource assessment. *Jour. Geol. Soc. India*, **66**: 713–724.

Sarkar, B.C., O'Leary, J. and Mill, J.B. (1988). An integrated approach to geostatistical evaluation. *Mining Magazine, London*, **59(3)**: 199–206.

Sarkar, B.C., Nayak, V.K. and Rao, P.S. (1995a). Statistical modelling of exploration data for grade potential of the phosphorite deposit, Khatamba blocks, Jhabua district, M.P. *Jour. Geol. Soc. India*, **46(2)**: 139–147.

Sarkar, B.C. and Nair, A.M. (2002). A Geostatistical modelling approach to Gold mineralization at Hutti mine, Raichur district, Karnataka. *Jour. Geol. Soc. of India*, **60**: 639–648.

Sarkar, P. and Bandyopadhyay, K.C. (1994). Study of Chotanagpur Gneissic Complex in between the area from Kotsila to Sindri, Purulia district, W.B. Unpub. G.S.I. Report, F.S. 1992-1993.

Sarkar, S.C. (1970). A study of the mineralisation of radioactive elements in Singhbhum Shear Zone, Bihar. *Proc. Nat. Sci. Acad.*, **36A**: 246–261.

Sarkar, S.C., Dwivedi, K.K. and Das, A.K. (1995b). Rare earth deposits in India—An outline of their types, distribution, mineralogy-geochemistry and genesis. *Global Tectonics and Metallogeny*, **5(1 & 2)**: 53–61.

Sarkar, S.N. and Rai, K.L. (2002). Geochemical modelling of sulphide mineralisation in Mosaboni-Rakha Sector of Singhum Copper Belt (Jharkhand) with reference to ore-genesis and mine exploration. *In:* Computer Applications in Mineral Development and Water Resources Management (Eds. K.L. Rai, G.R. Sahu and P. Diwan). South Asian Association of Economic Geologists, pp. 79–93.

Saukov, A.A. (1946). Geochemistry of Mercury. Akad. Nauk. SSSR, Inst. Geol.

Sengupta, R., Khalil, S.M., Rakshit, S., Deb Roy, D.K., Sinha, J.K., Mitra, S.K., Majumdar, S., Raghav, S. and Bhattacharyya, S. (1990). Multimineral placer deposits in the inner shelf off Orissa coast. *Geol. Surv. India*, Spl. Pub., **29**: 135–143.

Sengupta, R., Deb Roy, D.K., Mitra, S.K. and Bhattacharyya, S. (1992). Heavy mineral placers off Orissa coast. *Marine Wing News Letter*, **8(1)**: 7–10.

Senthiappan, M., Abdullah, N.M., Kumaran, K., Shankar, J., Durairaj, U., Nambiar, A.R., Michael, G.P., Unnikrishnan, E. and Ramchandran, A.V. (1987). Heavy mineral sands of Paravur-Varkala, Kerala. *Geol. Surv. India*, Spl. Pub., **24**: 215–225.

Sheppard, S.M.F. and Taylor, H.P. Jr. (1974). Hydrogen and oxygen evidence for the origin of water in the Boulder Batholith and the Butte ore deposit. *Montana. Econ. Geol.*, **69**: 926–946.

Shipulin, F.K. et al. (1973). Some aspects of the problem of geochemical methods of prospecting for concealed mineralization. *Geochem. Explor.*, **2**: 193–225.

Siddique, H.N. (1967). Recent sediments of the Bay of Bengal. *Marine Geology*, **5**: 249–291.

Siddique, H.N. and Mallick, T.K. (1972). An analysis of the mineral distribution pattern in the recent shelf sediments off Mangalore, India. *Mar. Geol.*, **12(5):** 359–391.

Siddique, H.N., Gujjar, A.R., Hashami, N.H. and Vallangkar, A.R. (1984). Surficial mineral resources of the Indian Ocean. *Deep Sea Research*, **31:** 763–812.

Sinclair, A.J. (1974). Selection of thresholds in geochemical data using probability graphs. *Jour. Geochem. Explor.*, **3:** 129–149.

Singer, D.A., Cox, D.P. and Drew, L.J. (1975). Grade and tonnage relationships among copper deposits. U.S. Geol. Sur., Prof. Paper **907A:** A1–A11.

Smith, R.E., Fracter, H.M. and Mocokpos, P.G. (1980). Golden Grove Cu-Zn Deposit, Yilgarn Block, W.A. *In:* C.R.M. Butt and R.E. Smith (compilers & editors), Conceptual Models in Exploration Geochemistry – Australia. *Jour. Geochem. Expl.*, **12:** 195–199.

Talapatra, A.K. (1966). Study of a part of Singhbhum granite around Jhinkpani, Singhbhum District., Bihar. Quart. *Jour. Geol. Min. Met. Soc. Ind.*, **38(10):** 18–32.

Talapatra, A.K. (1968). Sulfide minerilization associated with migmatisation in the south eastern part of Singhbhum Shear Zone, Bihar, India. *Econ. Geol.* (USA), **63:**156–168.

Talapatra, A.K. (1979). Gossan geochemistry as a guide to exploration in parts of Rajasthan, Western India. Proc. 7th Int. Geochem. Expl. Symp., pp. 173–184.

Talapatra, A.K. (1991). Case studies of mineral belt modelling applied to some base metal occurrences of India. Regional Workshop on Mineral Deposit Modelling, Dhanbad.

Talapatra, A.K. (1994). Recent trend in base metal exploration of concealed deposits with reference to Indian subcontinent. Bhu-Vidya Golden Jubilee Volume, 52–59.

Talapatra, A.K. (1999). Heavy mineral placer deposits of India with special reference to 3D Modelling in Parts of offshore areas for resources evaluation – suggestion. *Ind. Jour. of Geol.*, **71(1& 2):** 105–115.

Talapatra, A.K. (2001). A scheme of computer-based mineral deposit modelling with special reference to Precambrian resources evaluation. *Jour. Geol. Soc. Ind.*, **57:** 231–237.

Talapatra, A.K. (2004). Lineament controlled metallogenesis in parts of Chotanagpur Gneissic Complex in Northern part of Purulia district, West Bengal. *Recent Researches in Geology*, **17:** 101–107.

Talapatra, A.K. (2006). Exploration of concealed land and offshore deposits using mineral deposit modelling. Abstract Volume, 3rd Madhya Pradesh Science Congress, Bhopal.

Talapatra, A.K. (2014). Rare earth, rare metal and other important mineral resources of Purulia-Bankura area, West Bengal – Some new approaches. *MGMI Transaction*, **110:** 32–39.

Talapatra, A.K. and Bose, S.S. (1978). A review of nickel sulphide and associated base metal mineralisation in greenstone terraines of Western Australia and South India with emphasis on exploration. *Ind. Minerals*, **32(4):** 10–20.

Talapatra, A.K. and Bose, B.B. (1979). Mercurometric survey technique for exploration of concealed base metal sulphide mineralisation in parts of Rajasthan. *Ind. Jour. of Earth Sciences*, **6(2):** 162–174.

Talapatra, A.K. and Banerjee, P.P. (2002). Computer based modelling for exploration of concealed mineral deposits – Some Indian examples. *Ind. Jour. of Geology*, **74(1–4):** 291–306.

Talapatra, A.K., Bose, S.S. and Venkagi, K. (1981). Sulphur dioxide soil gas sampling for exploration of concealed sulphide mineralisation under sandy overburden. *Ind. Minerals*, **35:** 30–32.

Talapatra, A.K., Bhattacharyay, C. and Bose, S.S. (1984). Trace element distribution and its implication in the western extremity of the Singhbhum Shear Zone, Eastern India. G.S.I. Record, No. **114:** 77–88.

Talapatra, A. K., Mukhopadhyay, M.K. and Banerji, A. (1986a). Use of cluster analysis in ore deposit modelling of Pur-Banera-Bhinder mineralized belt, Rajasthan. *Ind. Minerals*, **40(4):** 11–21.

Talapatra, A.K., Talukdar, R.C. and De, P.K. (1986b). Electro-chemical technique for exploration of base metal sulphides. *Jour. Geochem. Expl.*, **25:** 389–396.

Talapatra, A.K., Bakshi, D. and Roy, S. (1991). Application of an ore deposit modelling technique in parts of Eastern India using qualitative data from known mineral belts. *Ind. Jour. of Earth. Sci.*, **18:** 71–83.

Talapatra, A.K., Sarkar Pradip and Bandopadhyay, K.P. (1995). Rare earth and rare metal bearing pegmatites along some lineaments, north of Purulia, West Bengal. *Ind. Journal of Earth Sci.*, **22:** 13–20.

Talukdar, R.C., Talapatra, A.K. and De, P.K. (1985). Application of field electrochemical technique for base metal exploration. Proc. 72nd Ind. Sci. Cong. Assoc. (Abstract Volume), part III: 73–74.

Taylor, G.F., Wilmshurst, J.R., Butt, C.R.M. and Smith, R.E. (1980). Gossans and ironstone. *In:* C.R.M. Butt and R.E. Smith (Eds), Conceptual models in exploration geochemistry, Australia. *Jour. Geochem. Expl.*, **12:** 118–122.

Taylor, G.P. (1979). Pathfinder element geochemistry in base metal exploration, North-West Queensland. *In:* J.E.G. Lover, D.I. Groves and R.E. Smith (Eds), Pathfinder and Multi-element Geochemistry in Mineral Exploration. Univ. of W.A. Extension Services.

Teertstra, G.K., Cerny, P. and Hawthrone, F.C. (1998). Rubidium Feldspars in granitic pegmatites. *Canadian Mineralogist*, **36(2):** 483–496.

Ure, A.M. (1975). The determination of mercury by nonflame atomic absorption and fluorescence spectrometry – A review. *Anal. Chim. Acta.*, **76:** 1–26.

Ure, A.M. and Shand, C.A. (1974). The determination of mercury in soils and related materials by cold atomic absorption spectrometry. *Anal. Chim. Acta.*, **72:** 63–77.

Webb, J.S. and Howarth, R.J. (1979). Regional Geochemical Mapping. Transaction Royal Society London, pp. 81–93.

Webb, J.S., Fortescue, J., Nichol, I. and Tooms, J.S. (1964). Regional geochemical reconnaissance in the Namwala concession area, Zambia. Tech. Comm. 47. Geochem. Pros. Res. Cen., Imperial College, London, and Geochemical Maps Nos 1-10 published by the Geol. Surv. of Zambia, 1964.

Webb, J.S. and Atkinson, W.J. (1965). Regional geochemical reconnaissance applied to some agricultural problems in Co. Limeric, Eire. *Nature, London*, **208:** 1056–1059.

Webb, J.S., Nichol, I. and Thornton, I. (1968). Broadening scope of regional geochemical reconnaissance. Proc. 23rd Int. Geol. Cong. (Prague), **6:** 136–147.

Webb, J.S., Thornton, I., Thompson, M., Howarth, R.J. and Lowenstein, P.L. (1978). The Wolfson Geochemical Atlas of England and Wales. Oxford University Press, p. 70.

Windley, B.F. (1984). The evolving continents. John Wiley & Sons. New York, p. 399.

Zipf, G.P. (1949). Human behavior and the principle of least efforts. Hafner Publishing Co. New York, p. 573.

Index

A

Air-borne
  geophysical data, 173
  hyperspectral data, 172
Anomaly, 66
Aravalli Fold belt, 16
Archaean granite-greenstone evolutionary
      sequence proterozoic geologic
      history, 16
Archaean plate tectonic model, 11, 12
Atomic Absorption Spectrometry/
      Spectroscopy (AAS), 127, 167
Atomic Minerals Directorate (AMD),
      Govt. of India, 172

B

Banded iron formation (BIF), 11–13, 15
Beach placers
  common minerals found in, 5
  of India, 175
  of India and their offshore
      extensions, 36–41
  minerals found in, 43
  to and fro wave action of ripples, 5
  wind movements, 5
Bijawar belt, 16
Biogenous deposits, 33–34
  in India, calcareous deposits, 33

C

Calcret, 71
Cassiterite rich placer, 5
Central Indian shield, Precambrian geology
      of, 14, 15

Chemogenous deposits, 34–35
  bedded-type phosphate deposits, 35
  geologically recent phosphorites, 34
  geoscientists of Marine Wing, GSI, 34
  phosphates, 34
  phosphorites, 34
Classical statistical analysis, 4
Classical statistics, 4
Cluster analysis, 18, 22, 113, 154, 160
Compressional arcs, 58
Concealed mineral deposits, non-conventional
      techniques, 117–143
  Aladahalli Belt, geology of, 137
  CHIM-10, 135
  field electrochemical technique, 134–140
  field sampling method, 138–140
  vapour geochemistry, application
      of, 118–134
Continental margin, deposits of, 31–35
  biogenous deposits, 33–34
  chemogenous deposits, 34–35
  diamond placers of Namibia (south west
      Africa), 31–33
  geomorphic barriers and traps, 33
  placer deposits on modern beaches, 32
    rich cassiterite placers of Myanmar-
      Thailand-Malaysia, 32
  terrigenous deposits, 31–33
  terrigenous material, uninterrupted
      transport of, 32
Crescent-shaped Cuddapah basin, 15

D

Deep-seated dispersion, 64
Deposit appraisal, 25

© Capital Publishing Company, New Delhi, India 2020

A. K. Talapatra, *Geochemical Exploration and Modelling of Concealed Mineral
Deposits*, https://doi.org/10.1007/978-3-030-48756-0

Printed by Printforce, the Netherlands